HELPING YOUR CHILD AT HOME

...with Mathematics

Macmillan/McGraw-Hill
School Publishing Company

EDITORIAL AND PRODUCTION STAFF

Editorial Director: Catherine White
Executive Editor: Ruth B. Cochrane
Editorial Production Manager: Helen Mischka
Editors: Barbara A. Everett, Lisa K. Higgins, Patricia S. Haupt

All photos on cover courtesy of Witherspoon Design and Will Shively Studio.

Macmillan/McGraw-Hill School Division
866 Third Avenue
New York, New York 10022

Printed in the United States of America

Level 3 ISBN 0-675-13095-6

9 8 7 6 5 4 3

Dedication

Macmillan/McGraw-Hill School Publishing Company is pleased to dedicate these materials to the following people, who have been our inspiration:

For America's classroom teachers, who strive for excellence in education and who need as much support at home from parents as they can get . . .

For America's parents and guardians, who seek better ways to support and interact with their children . . .

And principally for America's schoolchildren, who face a future filled with wondrous opportunity and serious challenge, and who need all of the educational resources we can provide as they carry our nation's hope and aspirations into the future.

Foreword

A MESSAGE TO PARENTS AND GUARDIANS

This may well be the single most important book you've ever used. By spending 10 minutes a day with your child doing the activities in this book—you can make a difference in your child's ability to learn more, learn faster, and get better grades.

There is indisputable evidence that children whose parents or guardians help them at home and stay in touch with their classroom teachers score higher than children whose parents are not involved.

- You don't have to have a high school diploma or college degree to help.
- It doesn't matter if you're rich, poor, or somewhere in between.
- It doesn't matter whether you are a white parent or guardian, a minority parent or guardian, or a single parent.
- It doesn't matter whether you live in a big city, a suburb, a small town, or a rural area.

IF YOU ARE WILLING TO COMMUNICATE WITH YOUR CHILD'S CLASSROOM TEACHER AND SPEND ONLY MINUTES A DAY WITH HOMEWORK AND THE ACTIVITIES IN THIS BOOKLET—YOU CAN HELP YOUR CHILD DO BETTER IN SCHOOL.

Your Child's Future Is at Stake

There are thousands and thousands of dedicated classroom teachers and educators across the United States who strive to provide the best possible education for your son or daughter.

But TEACHERS and YOUR CHILD NEED YOUR HELP to succeed.

Your child's classroom teacher devotes an average of six hours a day guiding your child to learn. Like you, teachers want your son or daughter to be able to get a good job and earn a decent living when they leave school. Teachers also want to make sure our children have the academic skills they need if they want to go on to college.

You Can Make a Difference in Your Child's Education . . . . And It Takes Only About 10 Minutes a Day!

Macmillan/McGraw-Hill's *HELPING YOUR CHILD AT HOME...with Mathematics* home study activities are designed to improve your child's understanding of math, permit you to participate in the activities with comfort and ease, and foster communication between you and your child.

- Parents do not have to have knowledge of math to do the activities.
- Each activity takes only about 10 minutes to do.
- Parents and their child do the activities together.
- Each activity includes easy-to-follow, step-by-step directions.
- You don't have to buy any materials to do the activities.
- You and your child will have fun doing the activities.
- Each activity reinforces a math concept your child is learning in school.

Every Child Needs Math Skills

Besides being able to read and write, math is arguably the most important skill your child uses every day of his or her life.

From the very basics—learning how to count, to tell time, to read a calendar . . . then to add numbers, multiply and divide—your child needs math skills to survive. Later on, your child will learn higher level math skills that will enable him or her to solve complex problems, to reason, and to acquire skills that will enable your son or daughter to get a good job.

During these early years, you can help prepare your child for the future. As a youngster, when your child sees that you value the knowledge of math, he or she will value it too.

We urge you to do the activities in this booklet with your child. Talk often to your child's classroom teacher. And make sure your children do their homework. The rewards will be great.

Acknowledgements

The Macmillan/McGraw-Hill School Publishing Company's *HELPING YOUR CHILD AT HOME...with Mathematics* was conceived by Chauncey A. Cochran and Stephen G. Clapp and developed in cooperation with Macmillan/McGraw-Hill and Cochran/Clapp, Inc.

Learning Design Associates, Inc., of Columbus, Ohio, provided editorial and production services for the development of *HELPING YOUR CHILD AT HOME...with Mathematics*. We gratefully acknowledge the help of John Buterbaugh, Project Manager, and Benjamin Shriver, Production Editor.

EXECUTIVE PROJECT EDITORS
Stephen G. Clapp
Chauncey A. Cochran

MANAGING AUTHOR
Richard W. Dykes, Austin, Texas

LEAD AUTHORS
Barbara Forth, Austin, Texas
Elizabeth Heath, Austin, Texas
Suzanne Kelley, Austin, Texas
Marilyn Neil, Birmingham, Alabama
Dwight Galbi, Lake Oswego, Oregon
Andrew Baumgartner, Augusta, Georgia
Lauren Caywood, Augusta, Georgia
Kay LeRoy, Augusta, Georgia
Jeanne Stroebel, Augusta, Georgia

CONTRIBUTING AUTHORS

TEACHER	SCHOOL	LOCATION
Faye L. Adams	Seven Hills School	Owensboro, Ky
Joy Allen	Sudderth Elementary	Monahans, TX
Barbara J. Arends	New Hope Elementary	New Hope, AL
Pennie Ascowitz	Pinewood Elementary	N. Lauderdale, FL
Jackie Barnes	Eagle Heights Elementary	Azle, TX
Sarah M. Barranco	Fultondale Elementary	Fultondale, AL
Marcelline B. Barry	Jacksonville High	Jacksonville, FL
Patricia M. Basden	Goliad Elementary	Odessa, TX
Linda E. Beal	Taylor County High	Campbellsville, KY
Betty H. Benfield	A.G. Cox Middle School	Winterville, NC
Margery J. Bessler	Whittier Elementary	Cincinnati, OH
Bernice A. Best	Beechgrove Elementary	Independence, KY
Diane H. Boesch	Hurlbutt Elementary	Weston, CT
Ann Boisse	Santa Rita Elementary	San Angelo, TX
Mary E. Bolin	Lackawanna Elementary	Jacksonville, FL
Bruce A. Boling	Eastwood Middle School	Indianapolis, IN
Diana Brettschneider	Rockdale Elementary	Cincinnati, OH
Judy A. Brown	Pleasant Valley Middle	Brodheadsville, PA
Jane Campbell	Caney Elementary	Pippa Passes, KY
Carole H. Carroll	Skyland Elementary	Tuscaloosa, AL
Mary Christensen	Lamkin Elementary	Houston, TX
Jeannie Churchwell	Norwood School	Birmingham, AL
Nicholas A. Cipparone	Bardstown Middle School	Bardstown, KY
Robin-Lynn Clemmons	Holy Innocents' Episcopal	Atlanta, GA
Carolyn Conklin	Bess Race Elementary	Crowley, TX
Kay Cook	Alief Independent School	Alief, TX
Deborah D. Coons	Masonville Elementary	Utica, KY
Joyce Copeland	Elizabethtown State Vo-Tech	Elizabethtown, KY
Carol Coulter	Clearwater Elementary	Clearwater, KS
Maretta Countess	Ed White Middle School	Huntsville, AL
Bonnye Cox	Keystone Heights Elementary	Keystone Heights, FL
Odessa Jackson Craig	Riverchase Middle	Birmingham, AL
Mary A. Davis	Retired--Scarborough Middle	Mobile, AL
Regina P. Davis	Huntington Place Elementary	Northport, AL
Rebecca W. Dewitt	Eustis Middle School	Eustis, FL
Patricia C. Dickman	Oxford Junction Elementary	Oxford Junction, IA
Kathleen Eagleson	Trigg County Elementary	Cadiz, KY
Bonnie B. Ellerbee	Elem. Science Enrichment Ctr.	Griffin, GA
Melissa Elmer	Inter-city Baptist Elementary	Allen Park, MI
Marita H. Eng	Sandlewood Jr./Sr. High	Jacksonville, FL
George A. Erickson	Blessed Kateri Tekakwitha	Thoreau, NM

CONTRIBUTING AUTHORS

TEACHER	SCHOOL	LOCATION
Miriam M. Feinberg	Retired	Long Island City, NY
Dwayne Ferrell	Austin Peay State University	Clarksville, TN
Hope Friend	Wm. J. O'Shea Intermediate	New York, NY
Leigh Ann Gahn	Roberta B. Tully Elementary	Louisville, KY
Susan Goble	South Elementary	Brownwood, TX
Veronica Granger	Fulton County Elementary	Hickman, KY
Merdeces Guerra	T.M. Clark School	Portland, TX
Patsy Ann Gurganus	Reform Elementary	Reform, AL
Cina Guthrie	Woodrow Wilson Junior High	Dayton, TX
Karen E. Hampton	Wyoming Indian Elementary	Ethete, WY
Ann C. Haney	Southwest Elementary	Murray, KY
Mary D. Hankins	Highland School	Cedar Hill, TX
Jean Hanson	Andrews Elementary	Austin, TX
Carole Harris	Croton Elementary	Melbourne, FL
William R. Hastings	Retired	Berea, KY
Teresa Heath	Bethume Elementary	Haines City, FL
Terry Hefner	Sugar Loaf Elementary	Killeen, TX
Faye Hilgart	Blair-Taylor Middle	Taylor, WI
Patricia O. Hobson	Riverdale High	Riverdale, GA
Shirley A. Holbrook	Southern Elementary	Georgetown, KY
Michael Lane Hubbard	DeWitt Elementary	DeWitt, KY
Jennifer Huff	Glendale Elementary	Vero Beach, FL
Eileen Hughes	Pine Ridge Middle School	Naples, FL
Malanie Jennings	New Prairie Junior High	New Carlisle, IN
Arthur V. Johnson II	Nashua Senior High	Nashua, NH
Esther W. Johnson	Indian Trail Elementary	Louisville, KY
Sandra L. Johnston	Roland H. Lewis Elementary	Temple Terrace, FL
Janet Jolly	Erma Nash Elementary	Mansfield, TX
Eleanor R. Kane	Fishcreek Elementary	Stow, OH
Arnalynn Kellner	Stricklin Primary	Refugio, TX
Mary Rhonda Kendrick	Millard Elementary	Pikeville, KY
Jane Kenneth	Van Buren Junior High	Tampa, FL
Patrick D. Kent	Upper Arlington High	Upper Arlington, OH
Patricia K. Kessler	Quebec Heights Camas	Cincinnati, OH
Betty Koleilat	Meyer Elementary	Houston, TX
Rita S. Koontz	Radcliff Middle School	Radcliff, KY

CONTRIBUTING AUTHORS

TEACHER	SCHOOL	LOCATION
Beth Lazerick	Hebrew Day Institute	Rockville, MD
Jane Lechner	Evans Junior High	Ottumwa, IA
Rochelle Z. Lieb	Pinecrest Elementary	Immokalee, FL
Mary H. Lineberger	Woodbridge Elementary	Tampa, FL
Mary K. Longoria	Bruni Elementary	Laredo, TX
Jody Lowry	Lake Gibson Junior High	Lakeland, FL
Chicha P. Lynch	Capuchino High	San Bruno, CA
Marge Maginnis	Lincoln Ave. Elementary	Lakeland, FL
Jean M. Manzano	Brown Junior High	McAllen, TX
Hilda Martin-Wade	W.L. Smith Elementary	Petal, MS
Lois F. McMillan	Roger B. Chaffee Elementary	Huntsville, AL
Shelly H. McNanaro	Enterprise City Elementary	Enterprise, AL
Barbara J. Mead	Stevens Park	Dallas, TX
Faye Meads	Admiral Thomas H. Moorer	Eufaula, AL
Winnie Milan	Coronado Beach Elementary	New Smyrna Beach, FL
Diane Miles	Beulah High	Valley, AL
Becky Moore	Club Estates Elementary	Corpus Christi, TX
Ethel P. Morris	Skyland Elementary	Tuscaloosa, AL
Jean Morrow	Emporia State University	Emporia, KS
Peggy A. Neal	Rochambeau Middle School	Southbury, CT
Marilyn S. Neil	Retired--Jefferson City Elem.	Birmingham, AL
Sue Newman	Santa Anna Independent	Santa Anna, TX
Terri Orn	Bowie Elementary	Corsicana, TX
Robin Osborne	Greenville Elementary	Greenville, KY
Shirley Parker	Huntington Elementary	Huntington, TX
Beth Paslay	East Cliff Elementary	Portland, TX
Sarah W. Pass	Southeastern Elementary	Remlap, AL
Karen Pearce	Seabourn Elementary	Mesquite, TX
Kathie B. Peeples	Mountain Brook Junior High	Birmingham, AL
Florine Martin Perry	Retired--Mercedes ISD	Mercedes, TX
Michael Petrozzino	South Broward High	Hollywood, FL
Carol Rainbolt	Stephen F. Austin Elementary	Edna, TX
O.J. Robertson	Retired--Salem Elementary	Russell Springs, KY
Jan L. Robinson	Cooper Junior High	Buffalo Grove, IL
Vicki Rose	Crestwood Elementary	Crestwood, KY
William Rottmann	John Long Middle School	Grafton, WI

CONTRIBUTING AUTHORS

TEACHER	SCHOOL	LOCATION
Donna S. Salter	Evergreen City School	Evergreen, AL
Jill Salzer	Christa McAuliffe Elementary	Woodbridge, VA
Melanie Samuelson	Old Lawn/Robertson Elem.	Round Rock, TX
Lorna K. Brooks Sapp	Straub Elementary	Maysville, KY
Linda C. Schaftlein	Ruth Dunn Elementary	Louisville, KY
Bonnie Sexton	Gatesville Elementary	Gatesville, TX
Judith D. Slinker	Clayton Elementary	Clayton, AL
Harriette Smith	Palm Harbor Middle School	Palm Harbor, FL
Jonnie Smith	W.H. Burnett Elementary	Terrell, TX
Sarah C. Smith	Flour Bluff Elementary	Corpus Christi, TX
Connie Stallard	Evarts Elementary	Evarts, KY
Jimmy W. Stanley	Bankhead Middle School	Cordova, AL
Laura S. Stephens	Brewbaker Primary	Montgomery, AL
Pat Sullivan	Nob Hill Elementary	Ruidoso, NM
Doris Terry	Retired--Hatton Elementary	Town Creek, AL
Judith L. VanVoorhis	Muskingum College	New Concord, OH
Karen Veith	St. Columban Elementary	Loveland, OH
Susan Watson	Sparks Elementary	Pasadena, TX
Sherri Whisman	Powell County Middle School	Stanton, KY
Gerald White	Ohio State University	Columbus, OH
Brenda Whitsell	Summer Academy	Gallatin, TN
Louise E. Wilhite	Retired--Decatur City School	Moulton, AL
Carl B. Williams	Fl. Sch. for the Deaf and Blind	St. Augustine, FL
Deborah Williams	West Texas Elementary	Stinnett, TX
Pam Williams	Dixieland Elementary	Lakeland, FL
Bettye Wilson	Sally B. Elliott Elementary	Irving, TX
Karen Winn	Early Elementary	Early, TX
Judy Wolfsohn	J.W. Johnson Elementary	Hialeah, FL
Mary H. Wood	Boyd Buchanan School	Chattanooga, TN
Mary J. Wrenn	Saint Anthony School	Hawthorne, NJ
Ann Wyatt	Alexander Elementary	Duncanville, TX
Sara Yahr	Sally B. Elliott Elementary	Irving, TX
Cathleen M. Zucco	Le Moyne College	Syracuse, NY

GRADE 3

CONTENTS

Page

SUGGESTIONS FOR USING THIS BOOK WITH YOUR CHILD

Here are a few simple suggestions for how you can use the ***HELPING YOUR CHILD AT HOME… with Mathematics*** series to help your child improve his or her mathematics abilities, perform better on tests, and earn better grades.

- If possible, set aside a specific time each day for you and your child to do an activity. You won't need a big block of time—most activities can be completed in 10 to 15 minutes—but it is a good idea to try to schedule the activity for a time when you are not likely to be interrupted.
- Work through the book at a pace that is comfortable for you. Start slowly, with one or two activities per week. Then, as you and your child become more familiar with the way these home-learning activities work, increase to three or four activities per week.
- Your child will benefit most if you work from the beginning of this book to the end. But don't worry about doing every single activity in the book. It's perfectly acceptable to skip one now and then if you don't have the necessary materials handy or if you think another activity looks more interesting. (And feel free to change or add to an activity based on your own personal experience!)
- When each activity is over, ask your child to tell you what he or she has learned or discovered. Talking in this way will help your child develop important thinking and communications skills.
- Encourage other family members to participate in the activities with you and your child. You might even consider making one evening each week a "family project night." Involving other family members will help develop your child's ability to work cooperatively in groups—and the shared sense of discovery will be a benefit for everyone in the family.

Part 1

Understand Numbers and Money: it's easier than you may think

A note to parents

Maybe you've already had "the Discussion" with your child about that five-letter word—-money. Chances are your youngster is already aware of the things money can buy, such as video games and action figures. But it's necessary to learn more. In the next few weeks, your child will learn to add and subtract money, to solve word problems about money, and to find equal amounts of money in different ways.

You can help your child learn more about numbers and money by doing these simple activities.

- Allow your child to pay for a purchase and let him or her count the change.
- Pretend you are just learning about money. Have your child determine how much money you will need to save every month to purchase the car you've been dreaming about.
- Try to measure everything in one room of your home, including those items that may be longer, shorter, or heavier than others.

Now try the activities in Part 1.

COUNTING TO WIN!

You need: 1 bag of beans or popcorn kernels

You and your parent each take a handful of beans.

Take turns using your beans to show the boxed numbers.

Work as fast as you can, but work carefully.

Check each other's piles to make sure there are no mistakes.

1.	1 ten	6 ones	6.	0 tens	2 ones
2.	3 tens	1 one	7.	2 tens	3 ones
3.	2 tens	7 ones	8.	3 tens	4 ones
4.	2 tens	0 ones	9.	3 tens	8 ones
5.	1 ten	5 ones	10.	4 tens	4 ones

NUMBER FIND

The pictures on this page show some ones, tens, and hundreds blocks.

Look away from the page and pick a letter between *a* and *j*.

Tell your parent your letter.

Look back to the page. When your parent points to the blocks by the letter you picked, you must tell the number shown by the blocks.

Repeat the game 5 times. Each time, read the number from the blocks.

a. 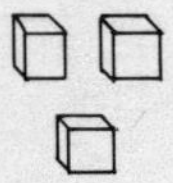

b. 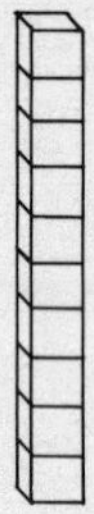

c. 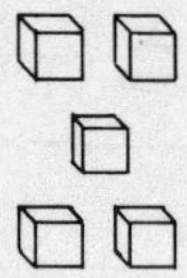

d. 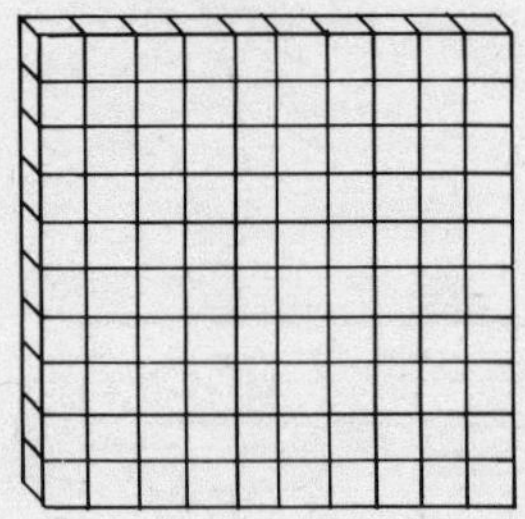

e. 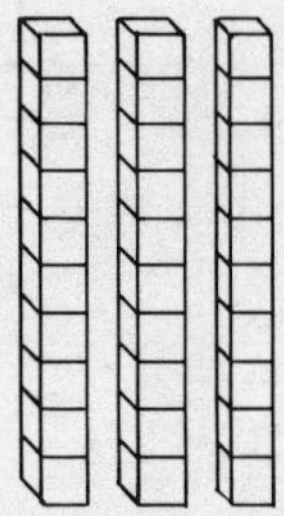

f. 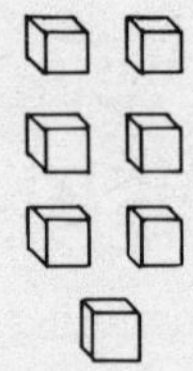

g. 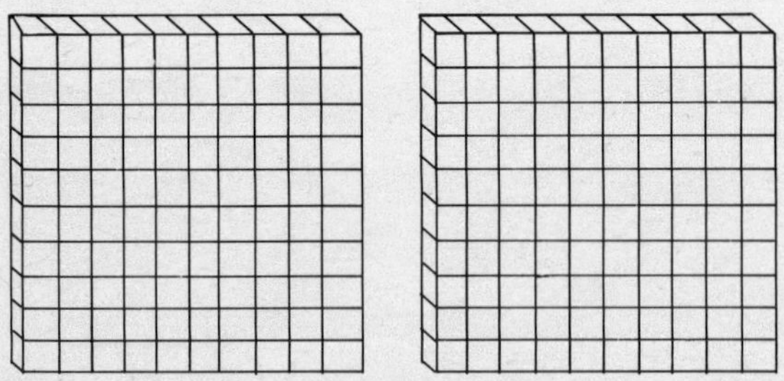

h. 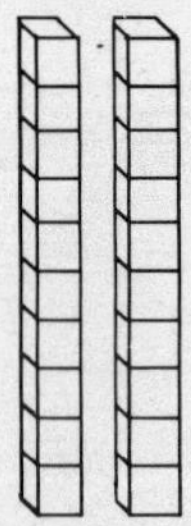

i. 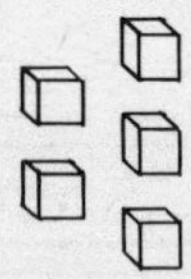

j. 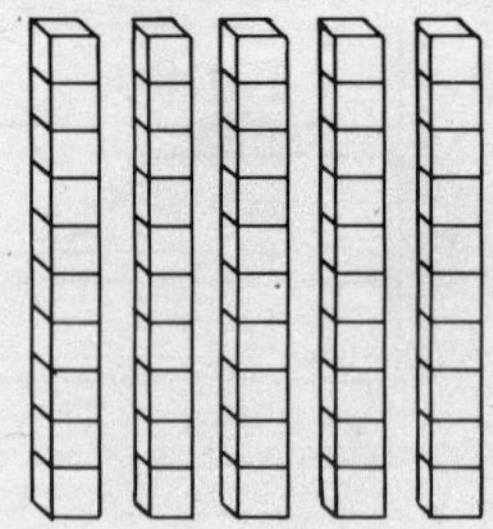

PICTURE MAKER

PICTURE MAKER
PICTURE MAKER

Bring along your imagination for this one!

Each of these shapes stands for a number.

Shapes:

ones tens hundreds thousands

To play:

- Look at each of the pictures below and decide what number it stands for.
- Explain to your parent how you came up with your answer.

1.

2.

3.

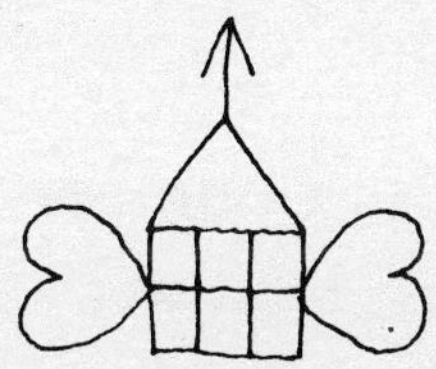

4.

5.

Can you draw a picture to show 2, 510? Try drawing some more big numbers.

COURAGEOUS COUNTERS

Take turns playing this number-building game with your parent:

- Choose 2 numbers from the bag.
- Place your numbers on the spaces below.
- The person who can make the greatest number wins the round and 1 point.
- Play until someone has 5 points.

Place number here	Place number here	O	O	O

If a 4 and a 7 were chosen, which number would you put first to make the greater number?

What would the number be?

You need: the cut-out numbers 1 through 9 from the back of the book, a bag

What is the least number you could make? ______

A RACE to the POT OF GOLD

You need: 2 game markers such as a coin or a button, number cards 1 through 5

Stack the numbers face down and put your markers on start.

To play:

- Each of you take a number. The higher number goes first.
- Take turns drawing a number card to move your marker. You must do what it says on each space. Miss a turn if you get it wrong!
- Return your number card to the bottom of the stack each time.

The first player to reach the pot of gold wins.

Finish 	Begin at 11. Count by ones to 18.	Finish the numbers. 500, 600, __, 800, __	The number __ comes after 36 when we count by ones.	Getting closer! Go ahead one.	Take an extra turn for being so smart.	Count by tens beginning with 30 and ending with 90.
Start Here	Begin at 50. Count by tens to 100.	Finish the numbers. 12, __ , __ 15, 16, __ , 18	You are a genius. Move ahead 2 spaces.	Which number comes before 29?	Begin at 400. Count by hundreds to 800.	Finish the numbers. 210, 220, __ , 240, __ , __ , 270

Fact Fun

You need: a bag of beans or buttons

Take turns reading the following stories and putting numbers into the problems. Use the numbers 1 through 9.

The other person counts out the answer with the beans.

For each correct answer you give, keep a bean. When you finish playing, count your beans to see who won.

1. _____ students were drawing a picture. _____ students finished. How many students were still working?
2. _____ monkeys were swinging on a vine. _____ monkeys fell off. How many monkeys were left?
3. There were _____ cats playing together. _____ cats went to dinner. How many cats were left?
4. _____ cows were grazing. _____ cows came and joined them. How many cows were there in all?
5. _____ children were going to school. _____ children were already there. How many children were there?
6. _____ students went to the party. _____ students stayed home. How many students were there in all?

Number Scramble

You need: game markers such as coins or buttons, cut-out numbers 1 through 8 from the back of this book

Place the cards face down and mix them.

Take turns choosing a number from 1 through 8.

Read aloud the question that goes with that number.

The first person to place a marker on the correct answer in the Answer Box wins the round. (Keep that number card if you win.)

1. Which is less, 38 or 82?
2. Which number comes right after 94?
3. Who has more grapes? Ben with 47 or Sally with 26?
4. Which number is one less than 652?
5. What is the least even number in the Answer Box?
6. Which is an odd number between 54 and 56?
7. Which even number comes between 24 and 28?
8. Which number comes after 99?

ANSWER BOX

38 2 82
651
95
76 47
26
31 77 55
100 93

Look at the Books

Each of you pick a book from below.

Put the numbers into the paper bag.

Take turns playing with your parent like this:

- Choose three numbers from the bag.
- Put them on your book in the same order that you chose them out of the bag.

The person with the greater number of pages in his or her book wins.

Play as many times as you like.

You need: cut-out numbers 1 through 9 from the back of the book, a small paper bag

Number of Pages in My Book

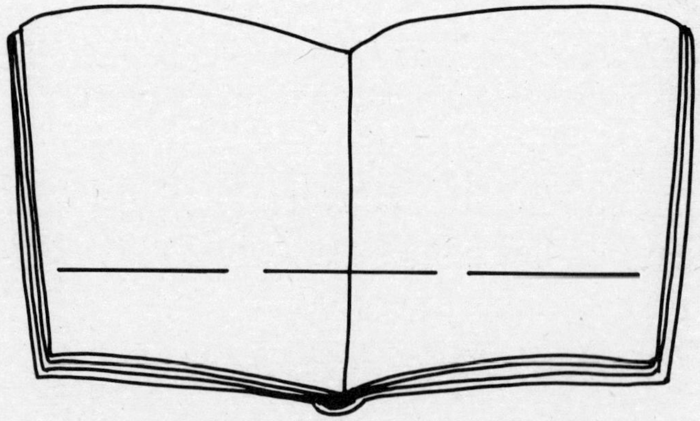

Number of Pages in My Parent's Book

HOW MANY ARE HERE?

HOW MANY ARE HERE?
HOW MANY ARE HERE?
HOW MANY ARE HERE?

You need: crayons or colored pencils

Look around your house. See how many of each of these you have: chairs, beds, pairs of your shoes, tables, children in the house.

Make a bar graph to show how many of each.

Color one square for each item you counted. Use different colors for each bar. Then you can compare the numbers.

chairs												
beds												
shoes												
tables												
children												
	1	2	3	4	5	6	7	8	9	10	11	12

Answer very quickly: What were there the most of?

Pattern Pictures

You need: green, red, and blue crayons

Count by ones. Write the missing numbers.

Color blue the numbers with 0 in the ones place.

Color red the numbers with 1 in the ones place.

Color green the numbers with 5 in the ones place.

Do you see a pattern?

Can you make your own pattern?

1	2	3	4			7			10
		13				17			
									30
					36				
								49	
				55					60
							68		
								79	
			84						
	92								

The Most Expensive

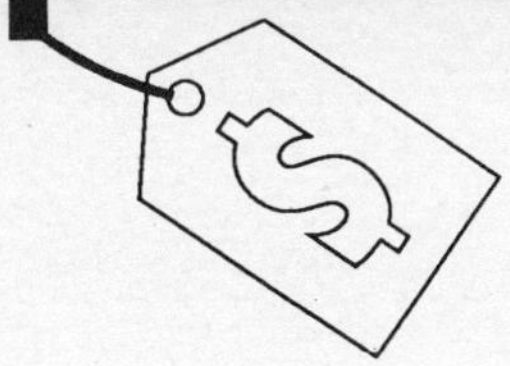

Extended Activity

Think of the greatest amount of money you can, and tell it to your parent.

Can your parent think of an even greater amount?

Is it possible to think of such a great amount of money no one could imagine a greater?

Which item in your home cost the greatest amount of money?

Think about the cost of different things.
Ask your parent if you could both visit an appliance store, a furniture store, and a jewelry store.

Make notes on the following page of the most expensive item in each store.

LISTING (The Most Expensive page 2)

Describe or draw the three most expensive items you found and their costs.

	Item		*Cost*
Electrical appliance		=	$
Furniture		=	$
Jewelry		=	$

Compare these prices on the next page.

CHARTING (The Most Expensive page 3)

Which item was the most expensive?

Draw a bar graph to display the values from the last page.

Ask your parent to help you choose a vertical scale which will cover all three amounts.

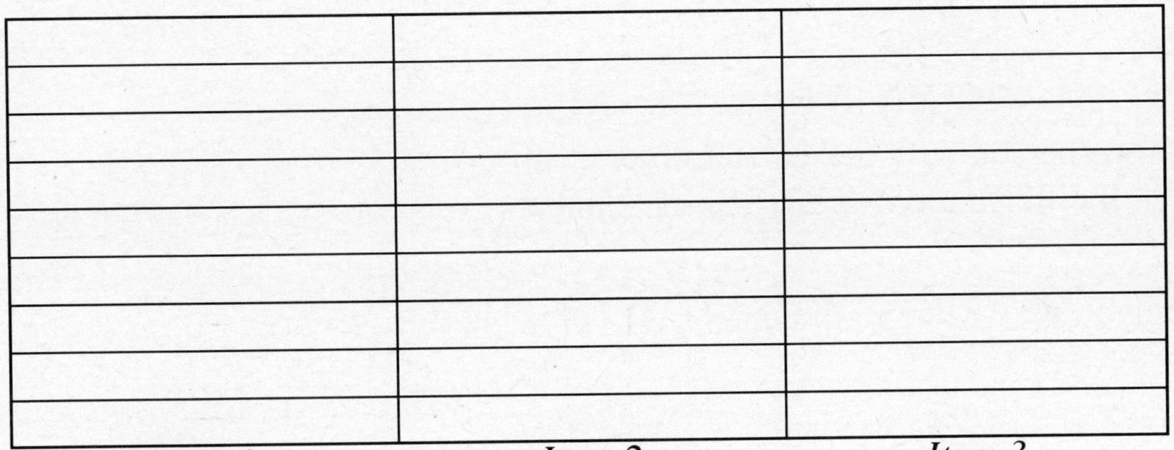

Item 1 *(Appliance)* | *Item 2* *(Furniture)* | *Item 3* *(Jewelry)*

You can easily find answers to the following questions from your bar graph:

1. Which item was the most expensive?
2. Which item was the least expensive?
3. What would be the total cost if you bought all three items?

A Bag of Riches

You need: a bag, cut-outs of coins from the back of this book

Put all the cut-out coins into the bag.

Take turns picking out 5 coins. Don't look when you pick.

Add up the coins you have chosen. The person with the greater amount of money on each turn wins a point. (In the illustration, the group of coins at right is the winner.)

Take a penny from the bag each time you win. (Put all the rest of the coins back every time.)

The person with the most pennies at the end of 5 games is the winner.

MONEY BAG

You need: a bag, coins from the cut-outs in the back of the book

Put the play coins in the bag.

Take turns playing this counting game:
- Take four coins from the bag.
- Count your coins to see how much you have.
- Use the coin counter below to help you count the money if you need to.

Play several times each.

COIN COUNTER
(Place your money here.)

______	______	______	______	______	______
Dollars	Half Dollars	Quarters	Dimes	Nickels	Pennies

Try playing a few times taking 6 coins from the bag.

Have your parent put little price tags on things in your home.

Go on a shopping trip and buy the things you like.

You must show your parent two ways you could give him or her the money to buy what you want.

Nothing should cost over $3.00.

You need: money from the cut-outs in the back of the book, paper, pencil or pen, tape

What did you buy that cost the most?

How much did it cost?

Funny Money

You need: cut-out money from the back of the book, a number cube, a bag

You and your parent will be putting on little skits.

Roll the number cube twice. Find the number from your first roll under A. Find the number from the second roll under B.

Read the story lines that go along with the numbers you rolled.

Act out the story problem by taking turns spending the money that is needed for the problem and figuring the change.

A.

1. Mary had $8.00 and bought a shirt for $4.20. She spent . . .
2. Jeffrey had $7.50 and purchased a hat for $4.77. He spent . . .
3. Karla gave Toni $3.25. Toni spent . . .
4. Brittney had $5.19. She spent . . .
5. Jay had $7.00 and bought a T-shirt for $4.10. He spent . . .
6. Out of her savings of $8.00, Kelly bought earrings for $3.16. She spent . . .

B.

1. $2.70 on ice cream. How much was left?
2. 89¢ on a notebook. How much was left?
3. $1.36 on a magic marker. What was left?
4. 74¢ on bananas. How much was left?
5. $2.52 on a flashlight. How much was left?
6. $1.80 on a paint set. How much was left?

MONEY IN YOUR POCKET

Lay out the money so both of you can reach it.

Take turns. At each turn:

- Roll the number cube and find the number you rolled in the list.
- Show the amount with the cut-outs.

Show each amount two different ways.

Amount List	
1.	$3.26
2.	$4.52
3.	$1.77
4.	$2.88
5.	$5.42
6.	$6.55

You need: bill and coin cut-outs from the back of the book, a number cube

Pick any one amount from this list and show it 3 different ways.

Part 2

To Add Or Subtract: the whole (number) story

A note to parents

Adding and subtracting are normal parts of everyday life. Whether you are totaling the grocery bill or figuring out how many of your 87 clothespins are still in the basket if 44 are on the line, you are using adding and subtracting. In reviewing the addition and subtraction facts in this section, your child will be preparing to solve such problems easily.

You can help your child use adding and subtracting in these everyday activities.

- Go over the use of doubles to calculate quickly. For example, your child can remember that 8 + 7 = 15 because he or she knows the doubles addition of 7 + 7 = 14 and 8 + 8 = 16.
- Write numbers out in word form, such as two hundred and thirty-six, explaining that check writing is one instance in which a person would write out a number rather than use the numbers themselves.
- Using a calculator, (because your math skills may be a little rusty) let your child see the difference between 624 + 257 and 426 + 752. Both use the same digits, yet each sum is different because of the different place value positions.

Now try the activities in Part 2.

Addition Action

You need:	2 markers (like a coin or button), pieces of paper numbered 1 to 5

You and your parent can play this game together.

Stack the numbers face down.

Play like this:

- Take turns drawing a number.
- Move your marker the correct number of spaces.
- Solve the problem in the space you land on. If you miss it, you must go back to the space you moved from.
- The first person to reach the finish line wins.

When you finish the game, play it again if you like.

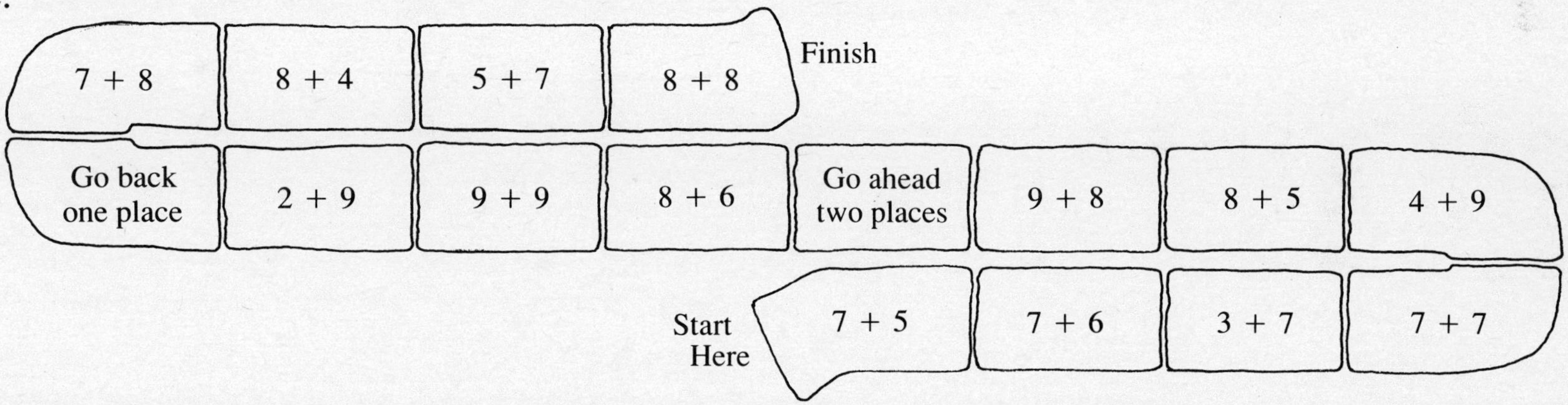

SUBTRACTION SEARCH SEARCH SEARCH SEARCH

You and your parent take turns finding subtraction facts in this puzzle.

The facts might run horizontally, vertically, or diagonally.

Decide how long you are allowed to search at each turn. A slow, silent count to 15 by the other player might be good.

You loop your facts with red. Your parent loops with blue.

When you have found all you can, count to see who had more.

You need: a red crayon and a blue crayon

10	4	6	1	3	2	5	8	7
2	2	3	8	1	8	4	1	3
8	9	7	6	2	5	10	6	4
7	5	2	4	0	8	9	1	4
7	10	1	7	3	0	4	2	6
0	8	10	2	9	1	10	5	1
1	1	3	6	7	2	5	3	0
6	3	9	2	5	4	4	8	7
4	4	1	5	5	9	9	7	2

Family Fun

Use the cards to form number sentences.

Show as many fact families as you can in 5 minutes. (The 4 problems on the list in the picture make up a fact family.)

As you show the number sentences with cards, your parent copies them on paper.

When time is up, count to see how many families you found.

Make up two word problems with any of your fact families and let your parent solve them. You check the answers.

You need: cards or paper squares with + , − , = , and the numbers 0 through 18

Macaroni Magic

Try this activity with your parent.

You need: pieces of paper numbered 1 to 20, a box of macaroni or some dry beans

- Your parent chooses a number from the pieces of paper.
- You take that many pieces of macaroni.
- Put the macaroni in groups to show <u>three</u> different numbers that add up to the number on the card.

  Here's an example:

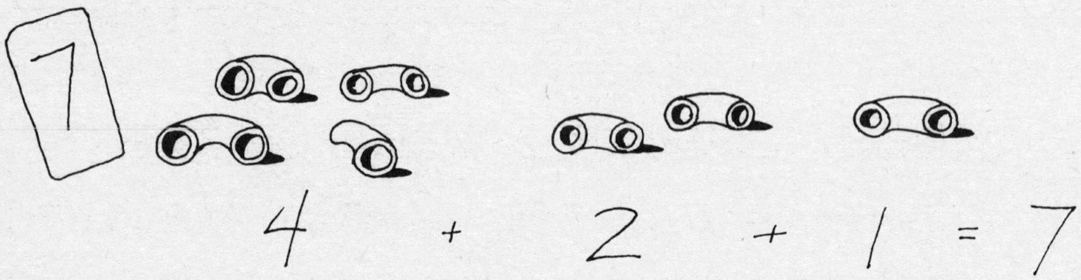

- Now you choose a number and let your parent try.
- Keep taking turns.

Can you form the shapes of some numbers out of macaroni?

Silly Stories

Have fun making up addition and subtraction math stories.

- Make up silly stories with your parent.
- Use the things in the Story Box.
- Make the story be a problem and have your parent solve it.
- Here is an example using 4 bears and 2 snakes.

There were 4 bears walking in the woods. They saw 2 snakes playing. All of them played together. The 4 bears invited the 2 snakes home for a delicious honey dinner. How many animals were there?

- Now have your parent make up a story and you solve it.

Story Box			
3 lions	5 books	1 chair	8 boxes
2 noses	4 jump ropes	9 elephants	4 kittens
7 shoes	6 monkeys	7 feet	6 rocks

Can you make up a story with six things from the Story Box?

Can you tell a story with eight?

How Many Will It Bee?

Take turns playing this game with your parent like this:

- Pick 2 numbers from the squares.
- Add them together.
- Draw that many bees in your beehive.
- Cross off the numbers after you use them.

  When all the numbers are crossed off, the game is over.

The player with more bees in a hive at the end is the winner.

Me

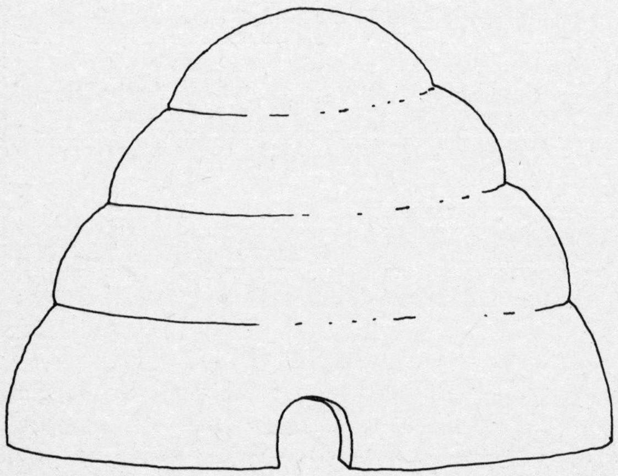

2	5	6
3	7	8
6	5	9
4	9	8

My Parent

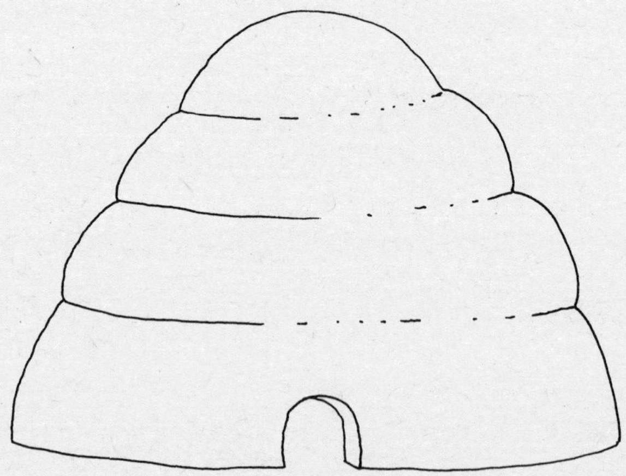

Did you come up with a good plan to win?

SMILING Subtraction

You and your parent can play this game.

You need: cards numbered 1 to 18, a small bag

Put the cards into the bag.

Then play like this:

- You and your parent each draw a card from the bag.
- Make a subtraction problem with the two cards.
- The person who drew the greater number must solve the problem.
- Each time you correctly solve a problem, draw one smiley face on your scorecard.
- Return the cards to the bag, after you've used them.

The first person to get 10 smiley faces wins.

My Score Card

Parent's Score Card

Can you subtract to find how many years you have to wait until you will be 14?

The TURN of the CARDS

You need: cards numbered 1 to 18

You and your parent are partners for this game.

Put the cards face down.

Choose two cards.

Your parent will tell you to add or to subtract.
Solve the problem.

Play the game until you have solved 5 addition and 5 subtraction problems.

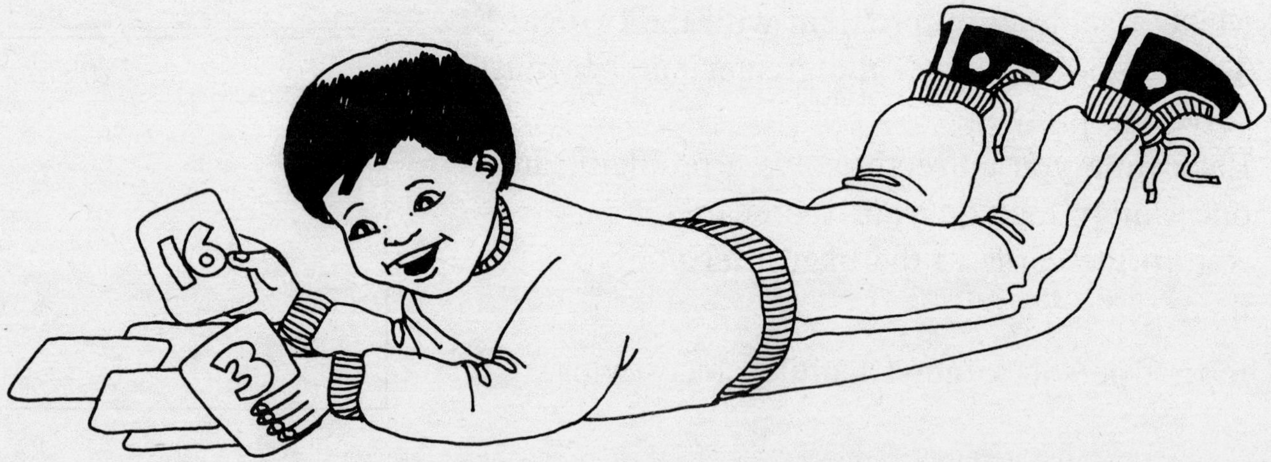

Can you give your parent some problems to solve?

Who Can Tell a Story?

You need: 2 markers, 5 pieces of paper numbered 1 to 5

Place your marker anywhere you like on the board.

Take turns with your parent.

At each turn:

- Draw a number. Move your marker that same number of spaces. You may move in either direction.
- Make up a word problem using the words where your marker started and where it ends. Your problem may be addition or subtraction.
- The other player must solve the problem to get a turn.

Play the game as long as you like.

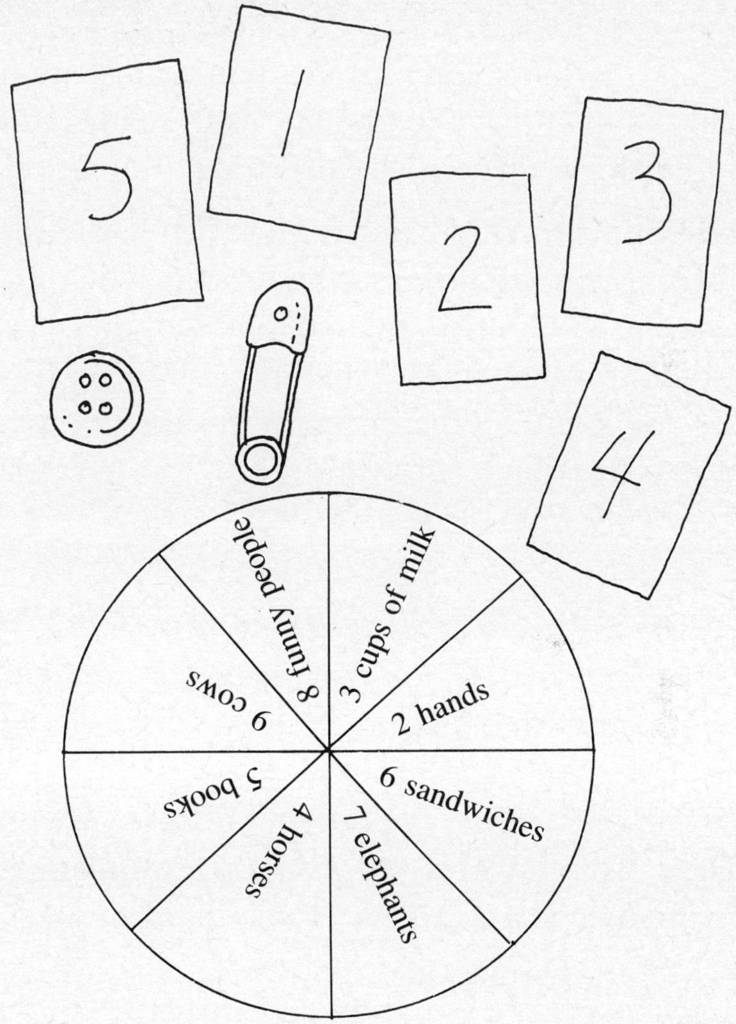

Can you think of a different game board you could make?

$upermarket $marts

Extended Activity

Take a trip to the supermarket with your parent. It makes a great place to practice your math. Be sure to take along a pencil and this book.

Does your supermarket sell more different kinds of cereal or more different kinds of dry dog food? Go to the aisle for cereal and count all the different cereal names. Then go to the dog food aisle and count all the different dry dog foods. Write your counts below.

_______ cereals

_______ dry dog foods

How many different kinds of bread does your store have for sale? Write the answer here _______. Are there more types of brown bread or of white bread? Count the breads. Subtract to find out how many more there are of one than the other.

_______ brown bread

_______ white bread

_______ the difference in breads

WATCH THOSE MISTAKES (Supermarket Smarts, page 2)

Imagine you can buy one of each kind of fruit for sale. Count all the different kinds of fruit. How many pieces of fruit will you get to take home?

Write your answer here ________.

Pretend there has been a mistake. Instead of one of each kind of fruit, you can buy one of each RED fruit in the store. How many pieces of fruit will you get to take home this time? Write your answer here ________.

How many pieces of fruit will you NOT be able to buy? (*Hint:* Subtract the red fruits from all the others.) Write your answer here ________ .

Oops! There's been a mistake again. You can buy one each of all the yellow and red fruit in the store, but no other colors. How many pieces of fruit can you buy?

_______ red fruits plus _______ yellow fruits = _______ pieces of fruit.

HARD PROBLEMS (Supermarket Smarts, page 3)

Now it's your turn. You have seen how easy it is to make up math problems in the supermarket. Now you and your parent each make up a good supermarket math problem for the other to solve.

Write a problem for your parent to solve here:

Have your parent write a problem for you to solve here:

What math problem do the cash registers work every time someone passes through the checkout line? ______________________________

____

How Many Beans Equal . . . ?

You need: a bag of dried beans

You can use beans to measure some things in your house. Lay the beans out beside these things. Write how many beans "long" each one is.

your hand __________ a cereal box __________

a pencil ____________ the television _________

a shoe _____________ a spoon ______________

Can you find 2 items that measure about 20 beans?

Can you find 2 items that measure about 30 beans?

How many beans long are the 2 longest items?

If you have a ruler, see how may beans equal an inch.

How could you measure your parent's height in beans?

Guessing Game

You need: a ruler with centimeters on it (cut out from the back of this book)

Try to find these 10 things in or around your home. Estimate the length in centimeters each would measure. Write your estimate next to the item below. Then actually measure to see how close you were. Write the real measurements by your estimates.

1. book
2. parent's thumb
3. pencil
4. window
5. door knob
6. pillow
7. drawer
8. mirror
9. fork
10. shoe

How many items were more than 30 cm?

Find 2 items that measure about 20 cm.

Find 2 items that measure about 35 cm.

TERRIFIC TENS

Here's an addition game to play with your parent.

You need: cut-out numbers 1 through 9 from the back of the book

Place the number cards face down. Mix them around.

Then, to play:

- Each player chooses a card.
- You put yours face up in front of the zero in the picture. Your parent's number goes face up after the one.
- The first player to give the correct sum of the numbers scores a point.

After 10 rounds, total the points to find a winner.

SCORE FORM		
	Me	*My Parent*
1		
2		
3		
4		
5		
6		
7		
8		
9		
10		
Total		

$$\begin{array}{r} \_\_\,0 \\ +\,1\,\_\_ \\ \hline \end{array}$$

What was the greatest sum you found?

Threes, Please!

- Take turns throwing the three coins onto the number chart. If you miss the chart, throw that coin again.
- Both players add their three numbers. See who has the greater sum each time.
- Play 5 times.

You need: three small coins or buttons, paper and pencil

Who got the greater sum most often?

9	17	18	13	86	75	28	62	45
59	15	28	43	26	31	97	36	80
20	24	7	37	16	45	83	35	88
92	31	27	47	26	11	6	55	32

MYSTERY SHAPE

Solve this one with your parent. Try to find which shape below we are describing.

1. The number in the shape is an even number.
2. The number in the shape is between 32 and 57.
3. The number in the shape is not the largest even number on this page.
4. The number in the shape is greater than 40.
5. The number in the shape is 2 more than 44.
6. Color the mystery shape.

Make up clues for one of the other shapes. See if your parent can find it.

Lost Mine Addition Adventure

Work out this addition story with your parent.

Pretend you find a lost diamond mine. It has five levels. You find one diamond at the first level. At the second level, you find one more diamond than at the first level. At the third level, you find one more than at the second level, and so on until you find diamonds at all five levels. First, see if you can add in your head all the diamonds found. Then write the numbers and add the diamonds to see how many you have:

First Level __________

Second Level __________

Third Level __________

Fourth Level __________

Fifth Level __________ What's your total? __________

Oh, no! On the way out, you drop your diamonds down a deep gorge. But then, taking a different turn, you come upon 28 new diamonds! Taking another new turn, you stumble over a pile of 73 more diamonds! If you can get safely out of the mine with these two groups of diamonds, how many will you have in all?

HOUSE of Bricks

Take turns building the House of Bricks. Begin at the bottom and work up the house. Add the two numbers with arrows between them. You fill in one answer in the empty brick and then let your parent fill in the next. See if every brick can be filled with the correct answer.

What were the two answers that completed the House of Bricks? ______ and ______

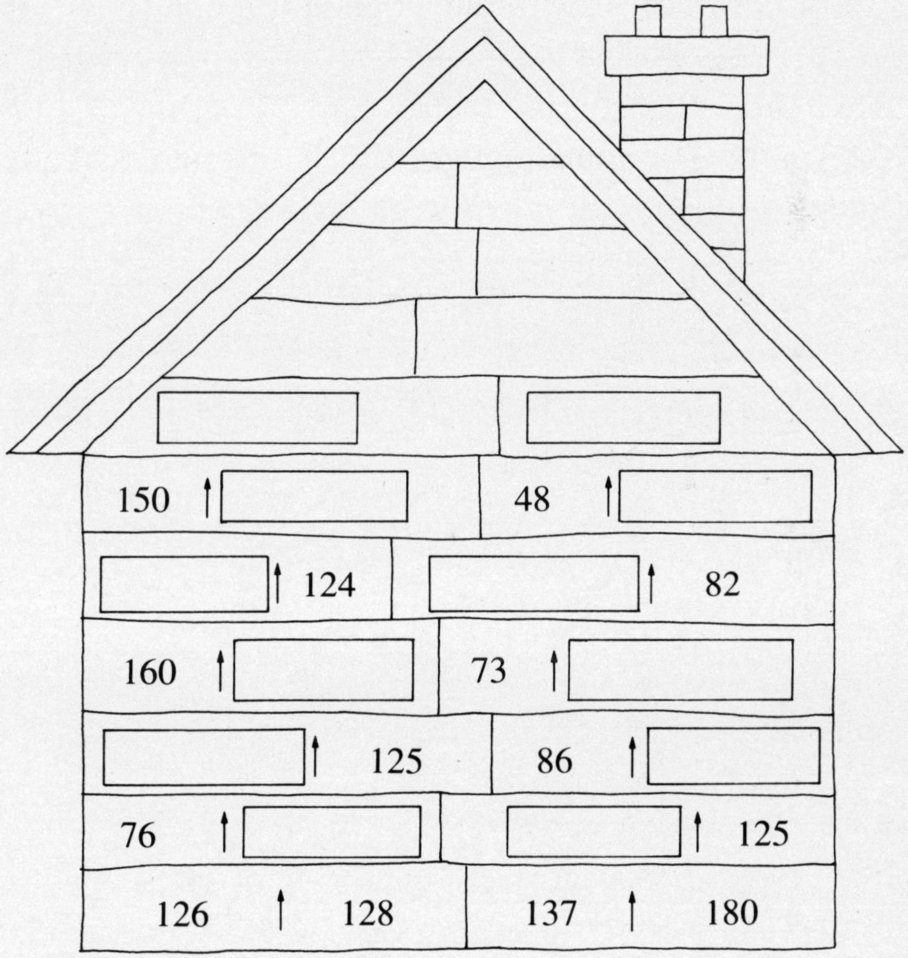

Estimation Station

You need: cards with the letters *A* through *G*

Take turns choosing two letter cards. Find two train cars that match the letters. Use rounding to estimate the sum of the numbers on the train cars.

Which two train cars would add up to the greatest number?

SHOPPING TRIP

You need: bill and coin cut-outs from the back of the book.

Take this shopping trip with your parent.

Decide who will be the buyer and who will be the seller.

The buyer chooses two items to buy. The seller adds the prices of the items and tells the buyer the total cost.

Use the money cut-outs to pay the bill.

Trade roles and play again. The buyer becomes the seller and the seller becomes the buyer.

Which two items cost the most?

SMART CHARTS

You need: popcorn kernels or beans

- Read the problems and look at the charts. Use the popcorn kernels for counting to find the numbers that go in the charts.
- Finish filling in the numbers in the charts to solve the problems.

Bobby and Susie sell lemonade for 5¢ a glass. If they sell 10 glasses, how much money will they make?

Number of glasses	1	2	3	4	5	6	7	8	9	10
Amount of money	5¢	10¢	15¢							

Karla and Jay sell old magazines for 8¢ each. How many magazines should they sell to make 64¢ ? If they sell 5 magazines, how much money will they make?

Number of magazines	1	2				6				10
Amount of money	8¢	16¢				48¢				80¢

Tossed Salad

You need: paper and pencil

Play this game with your parent.

- See who can make the cheaper salad. Choose three things to put in your salad. Write down the cost of each item and add them. Your parent does the same. Who has the cheaper salad?
- Play again to see who can make the more expensive salad with three things.
- Play the game again, but this time choose your three favorite vegetables. Whose was more expensive this time—yours or your parent's?

You could cut out pictures of your favorite salad fixings. Look in newspapers and magazines for pictures. Paste them on a piece of paper. If you cut the paper in an oval shape, it could be a salad bowl.

You're the Teacher Now

You need: paper and pencil,
18 beans or buttons,
scissors

Make 9 ten-blocks. Trace the ten-block below and cut it out. Make 8 more.

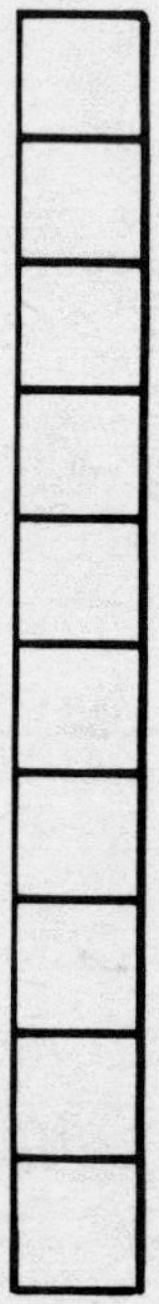

Take turns playing teacher with your parent.

- First, you be the teacher and your parent will be the student. Use the ten-blocks and beans or buttons to explain how to use renaming to solve the first problem below.
- Then, let your parent explain the next problem.
- Take turns explaining all the problems.
- Sometimes it's fun to make a mistake on purpose. See if your parent can catch you! But be careful, your parent might try to trick you, too!

27	43	58	39
+18	+19	+3	+37
68	87	73	66
+19	+6	+17	+27

HOW MUCH WILL WE NEED?

Extended Activity

Go to the grocery store with your parent.

Take along a pencil and the following page.

On that page there are several items listed.

Find the grocery store items listed (they do not have to be a certain brand) and write the prices in the price column.

When you get home, add the prices of the items shown on the chart.

Then complete the rest of the activities on that page and the following one.

SHOPPING LIST (How Much Will We Need?, page 2)

Add the prices of these items.

Item	*Price*
Bread	
Milk	
Dog food	
Lettuce	
Ketchup	
Potatoes	
Hamburger	
Tomatoes	
Total	

If you bought all the items, what would the total be?

If you have only $8.00, could you buy all the items? If not, what items could you leave off the list to keep your total below $8.00?

SHOPPING LESSONS (How Much Will We Need?, page 3)

Tell what you learned about the different prices of grocery items. Explain how addition and subtraction help you at the grocery store.

Button Bounces

Each player uses 2 markers.

Take turns dropping your markers onto the game board.

Subtract the two numbers on which your markers fall. Remember to subtract the lesser number from the greater one. The difference is your score.

Play at least 5 times. The higher score wins.

Who bounced onto the greatest number?

You need: 4 game markers, such as buttons or beans, paper and pencil

88	52	27	71	56	62	53	44	43	9
22	37	10	78	31	19	81	17	60	32
20	48	96	46	87	15	31	22	21	59
55	21	54	7	81	16	97	47	70	40

What Will My Number Be?

You need: a number cube and 2 game markers

Here's how to play:

- Your parent rolls the number cube.
- You place a marker in Box A on the little circled number that matches the one on the number cube.
- Your parent rolls again.
- This time you place a marker in Box B.
- Both of you subtract the two numbers.
- The first one to get the correct answer wins the round.

Play for about 10 minutes.
Who had more correct answers?

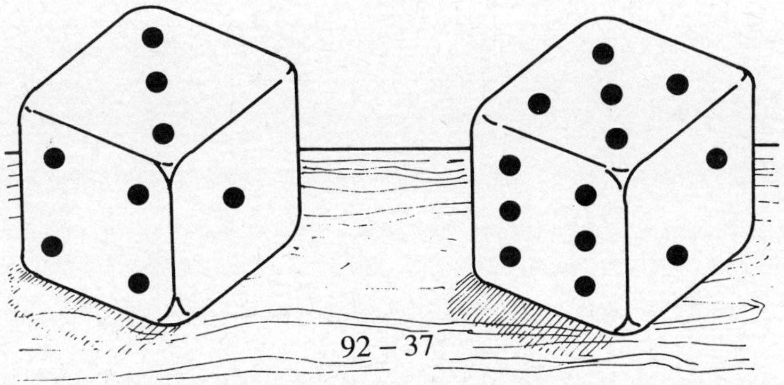

Box A

① 81	② 80	③ 92
④ 72	⑤ 53	⑥ 62

Box B

① 24	② 16	③ 28
④ 45	⑤ 37	⑥ 25

BOXED OUT

Take turns with your parent drawing a line to connect 2 dots.

Before long, the lines will start connecting to form boxes.

Each player starts with a score of 700. Every time you draw a line that completes a box, you get to subtract that number from your score.

Check each other's subtraction.

The first player to use up all 700 points—or whoever has the lesser number left when all dots are connected—wins.

78	56	45	21	62	47	19
37	22	85	33	37	64	90
83	46	74	57	44	67	26
52	84	28	63	77	35	42
43	75	34	25	86	36	27
54	87	65	47	53	76	24
73	49	55	72	66	32	82

Animal Funnies

You need: small papers or cards numbered from 1 to 10

Place the numbers face down.

You choose a number. Then your parent takes one.

Find the story parts to match your numbers in the box below, and together make up a story problem.

Make some addition problems and some subtraction. They can be silly!

Take turns solving and checking the problems.

1. 363 bears	6. 560 ducks
2. 156 owls	7. 421 frogs
3. 232 birds	8. 382 elephants
4. 485 snakes	9. 519 monkeys
5. 791 turtles	10. 440 giraffes

Choose 3 numbers. Make up a problem where you must add and subtract.

Climb the Ladder

Draw straws to see who goes first.

Start at the bottom of the ladder.

Take turns climbing up the ladder to find the differences.

Use the score key to find your score each time.

When you reach the top of the ladder, total the scores to see who won.

Score Key If your difference is:	
275 to 225	score 6 points
224 to 175	score 5 points
174 to 125	score 4 points
124 to 100	score 3 points
99 to 50	score 2 points
49 to 0	score 1 point

Turn	*Me*	*My Parent*
1		
2		
3		
4		
5		
Total		

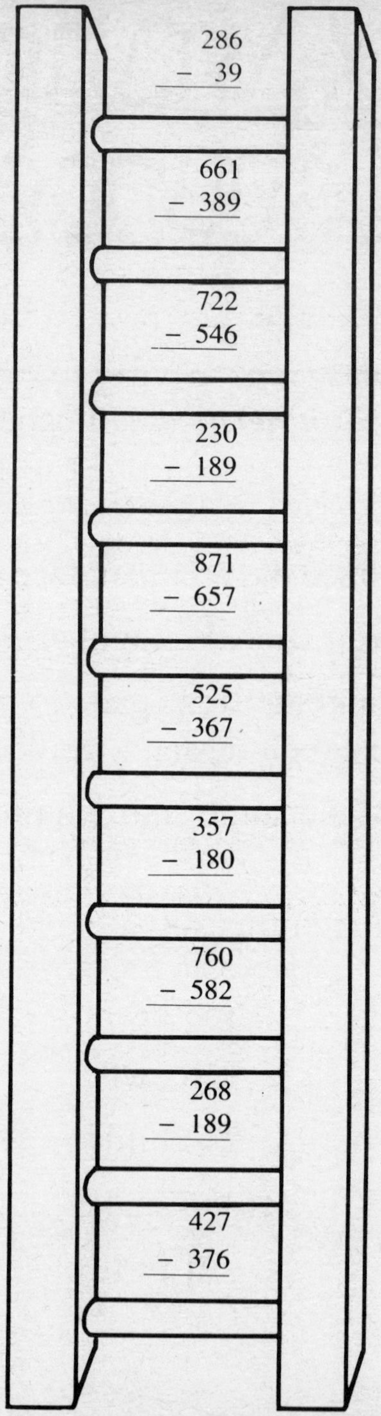

Subtracting on Target

You need: a bean or coin, paper and pencil

Play this game with your parent. You keep score on a sheet of paper.

Start the game with 300 points for each player.

Take turns dropping the bean onto the target.

After each drop, subtract the number where the bean landed from that player's score.

The player with more points left at the end of the game is the winner.

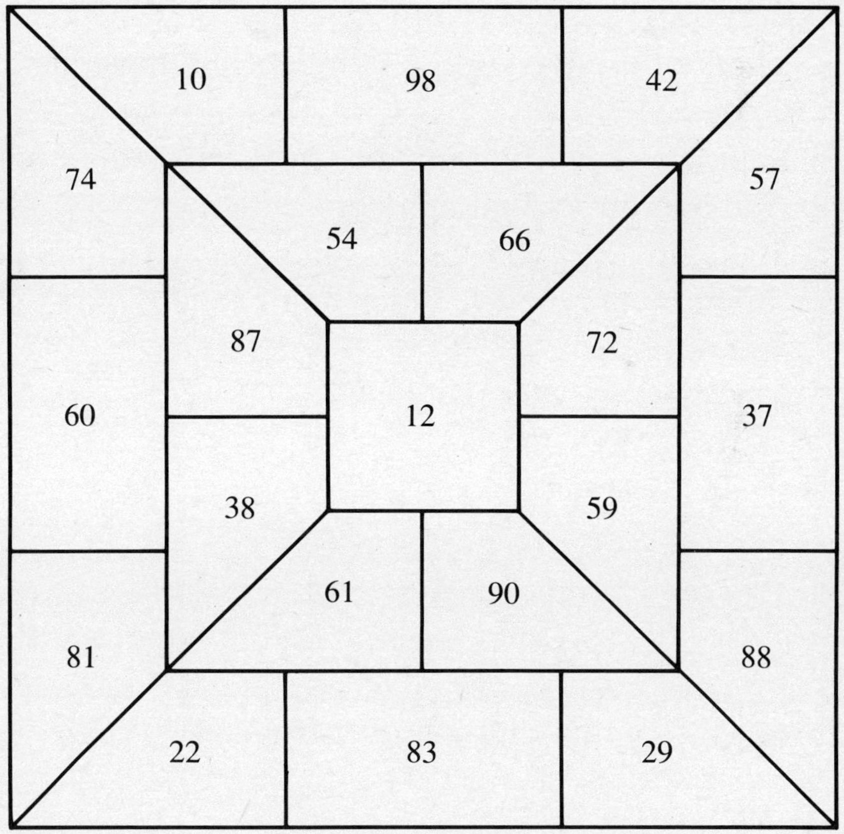

Think Back!
Think Back!
Think Back!

You need: small papers or cards numbered from 1 to 8

Take turns. Place the numbers face down. Choose a number and find it in the list below. Follow the directions for that number.

1. Make a word problem using 254 children and 129 parents. Subtract.
2. Think of three one-digit numbers and have the other player add them.
3. Subtract a one-digit number from a two-digit number.
4. Add two three-digit numbers.
5. Make up a story problem using 38 cats and 63 dogs. Add.
6. Count by hundreds. Which numbers are missing? 160, ____, 360, 460, ____
7. Complete the fact families:

8 + 4 = 12	7 + 8 = 15
[4] + [ ] = 12	[8] + [ ] = 15
12 − [ ] = [ ]	15 − [ ] = [ ]
12 − [ ] = [ ]	15 − [ ] = [ ]

8. Subtract a two-digit number from a two-digit number.

Pick and Score

Write the following digits, one per card:
0, 1, 2, 3, 4, 5, 6, 7, 8, 9, 9, 9.

Put the cards into a shaker.

Ask your parent to play the game with you.

You need: an index card cut into 12 small rectangles, a pencil, a shaker (like a box)

Each of you starts out with a score of 400. The object of the game is to finish with the lower total. Play like this:

- Pick two digits from the shaker without looking at their values.
- Arrange the digits to form a number in any order you like.
- For Turn 1, subtract your two-digit number from 400, and enter the difference on the scoreboard.
- For Turns 2 and 3, subtract the new two-digit number from the previous difference, and enter the new difference on the scoreboard.

Scoreboard

Turn	*You*	*Your Parent*
1		
2		
3		

Which part of the game is skill, and which part is luck?

Subtraction Rollers

Take turns rolling the number cube.

After each roll, write your number in one of your number boxes.

When you have each made a 3-digit number, compare your numbers.

The player with the greater number subtracts. The other player checks.

You need: a number cube

My Numbers				*My Parent's Numbers*			
1.				1.			
2.				2.			
3.				3.			
4.				4.			

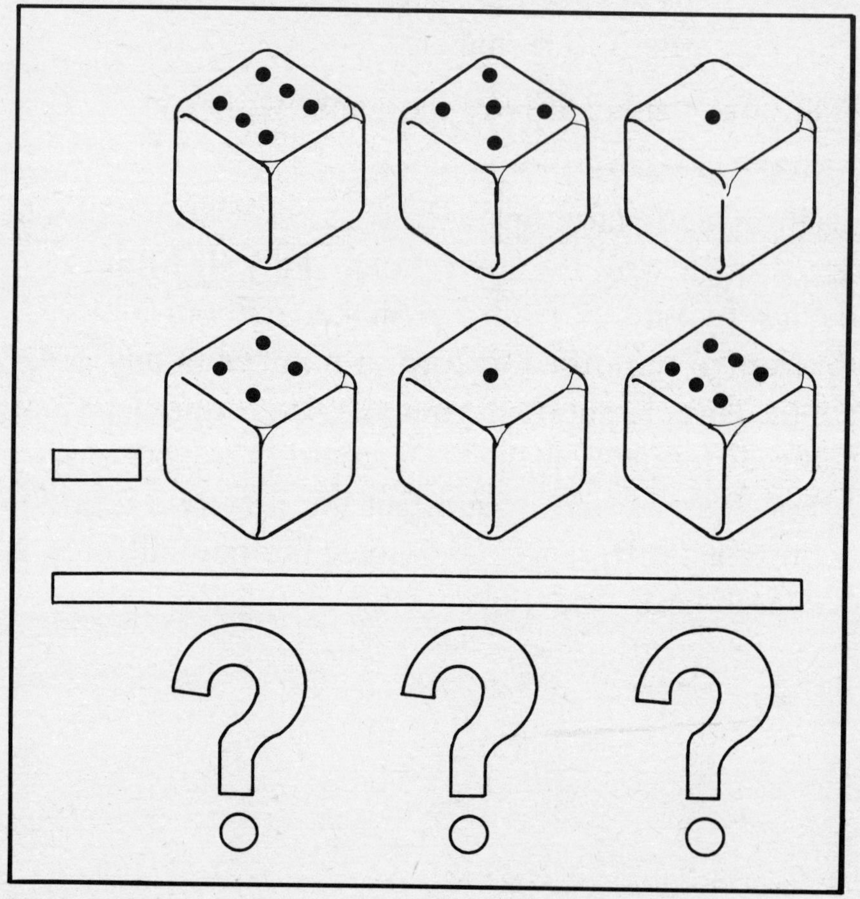

Subtraction Satisfaction

You need: paper and pencil, 2 buttons (different colors), a number cube

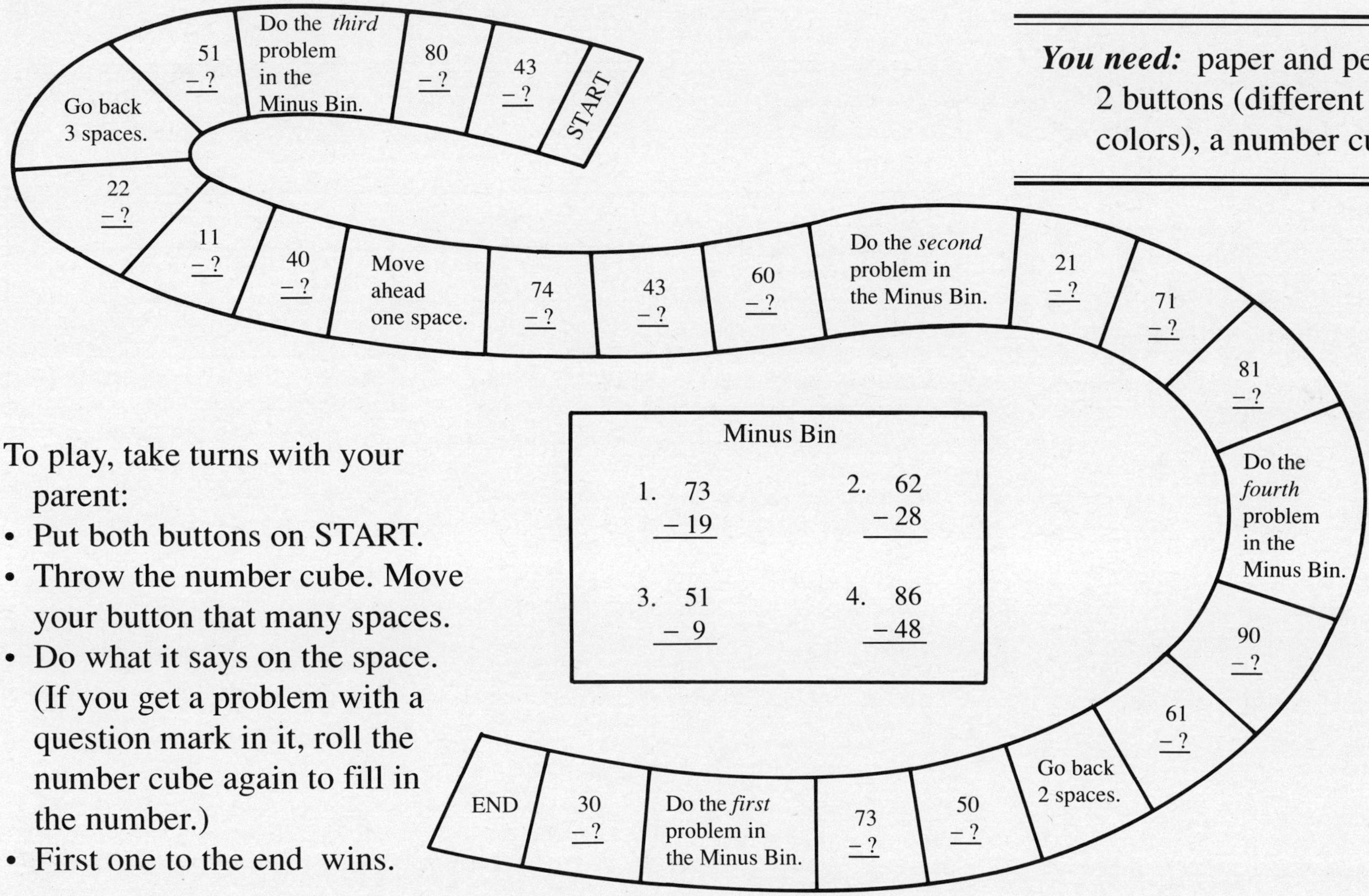

To play, take turns with your parent:

- Put both buttons on START.
- Throw the number cube. Move your button that many spaces.
- Do what it says on the space. (If you get a problem with a question mark in it, roll the number cube again to fill in the number.)
- First one to the end wins.

TALLY IT UP!

What to do:

1. Carry this page around the house. Count with your parent all the items listed in the chart.
2. Make a tally mark each time you count an item.
3. Then total the tally marks and write the total in the spaces provided.

	Tally	*Number of Tallies*
Pictures		
Doors		
Windows		
Tables		
Chairs		

Do you have more doors or windows?

Do you have more tables or chairs?

Now, make up your own question to be answered from the data.

Card-O

Draw the first 20 cards from a deck of cards. Put tally marks beside each card as it is drawn.

Then tell your parent the answers to these questions:

- What was the lowest card drawn?
- What was the highest card drawn?
- What card would be in the middle?
- What was the most frequently drawn card?

You need: a deck of cards, shuffled—let the ace equal 1

King	
Queen	
Jack	
10	
9	
8	
7	
6	
5	
4	
3	
2	
Ace	

Try it again with the next 20 cards. Are the answers to the questions the same this time?

I Want, I Want, I Want

Extended Activity

Make a list of 5 things you would like to buy if you could. It could be toys, books, clothes, or sports equipment.

With your parent, find out the prices of the 5 items. You could go to a store that sells the items on your list. You could call and ask for prices. You might find them in a newspaper ad or a catalog. Write down the price of each item.

Use the prices to do the activity on the next page.

MY "WANTS"

	Things	*Prices*
1.	____________________	__________
2.	____________________	__________
3.	____________________	__________
4.	____________________	__________
5.	____________________	__________

HARD CHOICES (I Want, I Want . . . , page 2)

With your parent's help, round each price to the nearest dollar. Use the rounded prices for this activity.

If you had $200 to spend on your “wants,” could you buy them all?

Would you have any money left over?

If you couldn't buy all your “wants” for $200, take something off your list. Subtract again. Now can you buy everything for $200 or less?

Is there any money left over?

What would you do with any leftover money: spend it or save it?

SPEND OR SAVE (I Want, I Want . . . , page 3)

Talk with your parent about spending and saving.

How do you decide what to spend your money on? How does your parent decide?

Do you ever save any money?

In a few sentences, write what you learned from this activity and the talk with your parent.

Part 3

Other Ways to Find Out How Much: multiplying, dividing and measuring time

A note to parents

In this hectic world, there are few certainties left. Paying taxes and solving math problems are among the very few things that don't change. Although you can't do much about your taxes, you can make math problems more fun for your child. In this section, besides reviewing the basics of measuring and telling time, your third grader can practice new skills in multiplication and division.

To make multiplying, dividing, and telling time more a part of day-to-day life, try some of these activities.

- Have your child tell you what time it is several minutes in a row. Then ask what time it was four hours ago.
- Have your child count money, first by counting pennies to see how many make a nickel. Then count how many nickels make a quarter and how many quarters make a dollar.
- If you buy a 12-pack of soft drinks, figure out together how many cans that would be for each family member.

Now have fun with the activities in Part 3.

Clock Work

You need: paper, scissors, a pencil

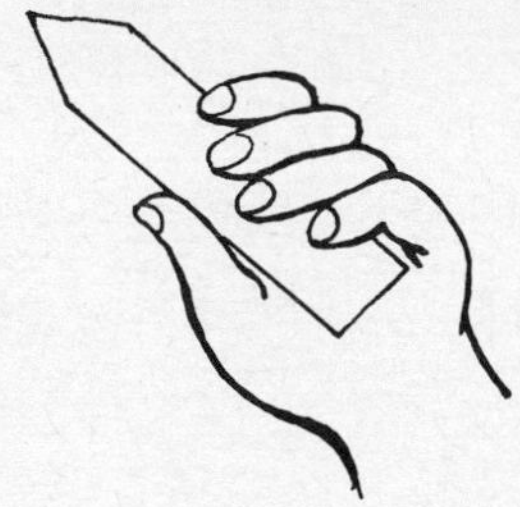

You and your parent make a clock. Use the circle on this page. Write the numbers on the clock correctly. Cut out a small and a large strip of paper for the hands. Make points at the ends. Place them on the clock and hold them in the middle. Take turns giving each other times to show on the clock. Be sure and say before or after the hour.

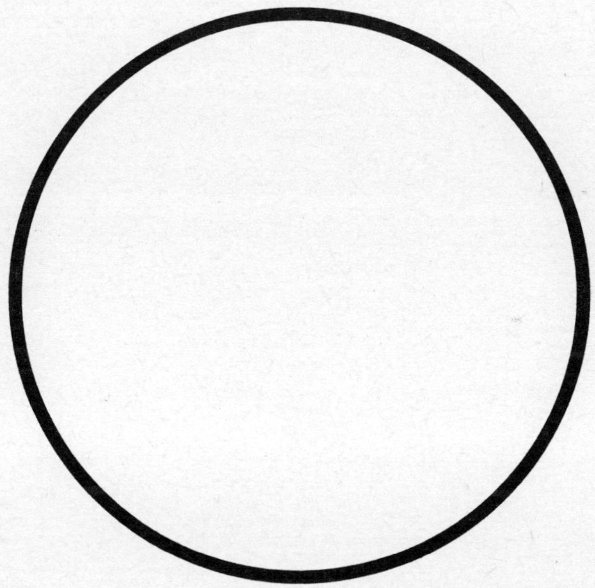

Show the time school starts on your clock.

TIME CLIMB

Each person places a marker at START. Take turns moving your marker to the next space. You must say the answer to the question correctly to move to the space. Whoever gets to the top first wins the time climb.

You need: 2 markers for game pieces (use buttons or beans), a clock

FINISH

30 minutes after 9:00 is __

20 minutes after 8:20 is __

50 minutes after 3:00 is __

It is 4:10. In 30 minutes it will be __

30 minutes before 7:30 is __

15 minutes after 12:25 is __

START

FINISH

20 minutes before 6:30 is __

It is 2:20. 20 minutes later is __

15 minutes before 2:45 is __

5 minutes after 2:20 is __

45 minutes after 6:05 is __

It is 3:35. In 12 minutes it will be __

START

What time is it 25 minutes after school is out?

How Long Does It Take?

Extended Activity

Time some of the things you do in a day. Try for 4 or 5 things. How long does it take you to eat breakfast, get dressed, play a game? Use the chart on the following page to record your times. If you go somewhere like the grocery store or ball practice, record the time that you leave and the time you return.

RECORDING EVERYTHING (How Long Does It Take?, Page 2)

My Time Chart			
What I did	*What time I began*	*What time I finished*	*Total time it took*

MY DAY WAS SO LONG (How Long Does It Take?, page 3)

Write a paragraph about your day. Include in your paragraph the times you recorded. How much time did you spend doing your favorite thing today?

THE GREAT BEAN RACE

You need: 2 markers (like coins or buttons), a number cube, a big bag of beans, 2 jars

- You and your parent put your markers at START.
- Take turns rolling the number cube.
- Move your marker the correct number of BEANS on the game board.
- Solve the problem in the BEAN, then check your answer by grouping real beans.
- If your answer is right, put the beans in your jar. If it isn't right, leave them out.
- Play until you reach the FINISH, or keep playing until someone's jar is full of beans. That's the winner!

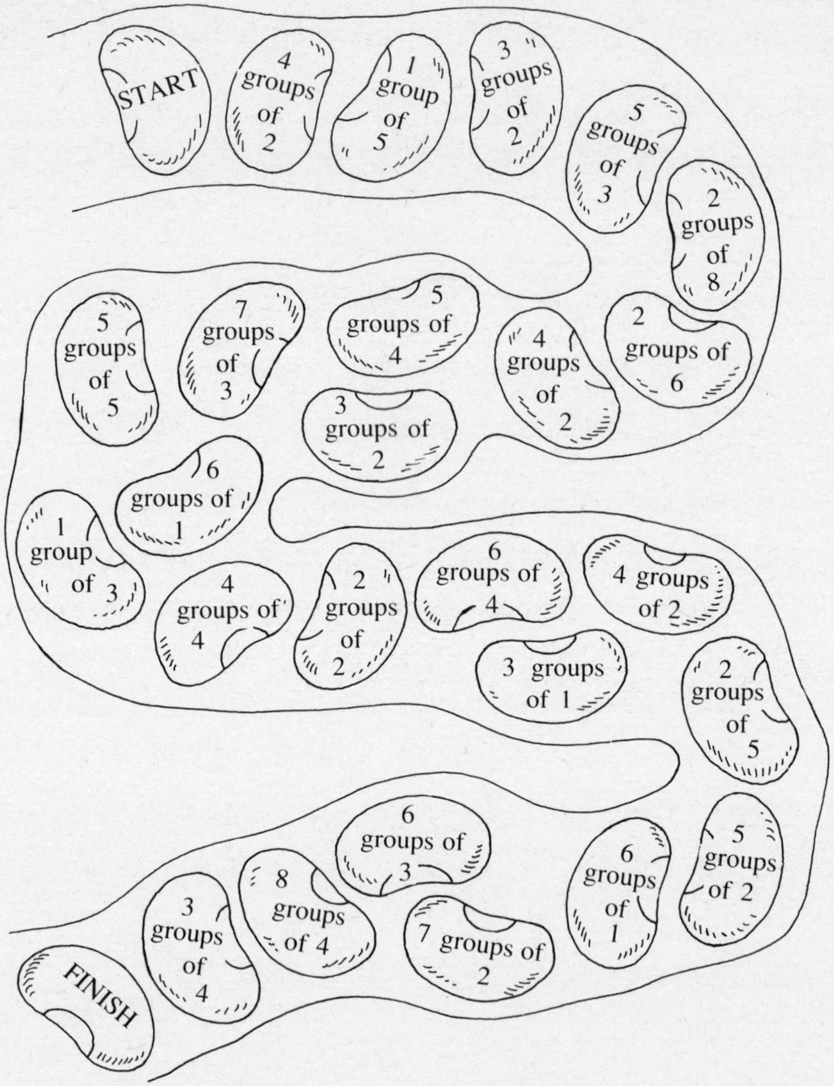

Visiting The Zoo

You need: 2 number cubes, a pencil, a paper clip

Hold the paper clip down at the center of the spinner with the pencil point. To play, then, take turns with your parent. (This time, let her or him go first.) At each turn:

1. Throw the 2 number cubes and spin the Zoo Wheel.
2. Make a multiplication sentence with the numbers and the zoo animal your pointer landed on. Example:
   - There are __5__ cages of __lions__ in this zoo.
   - Each cage has __4__ __lions__ in it.
   - So this zoo has __20__ __lions__ in it!
3. Keep spinning until every animal has been counted!

Draw a big map of your zoo. Draw the cages if you can.

The Zoo Wheel

Factor Bingo

Put the slips of paper in the bag and mix them up. Then, to play:

- You and your parent take turns picking a factor slip from the bag.
- Pick a number on your Bingo Card that is a product of the factor you picked and a missing factor.
- Make up a multiplication sentence with your two numbers. What is the missing factor?
- If you think of the missing factor, put a marker on the product square.

The first one to get four in a row—across, down, or diagonal—wins Factor Bingo!

You need: 12 markers (buttons, beans, etc.), a paper bag, 8 slips of paper marked 2–9 (factor slips)

My Bingo Card

4	27	15	8
12	10	32	9
6	21	16	18
24	14	40	35

My Parent's Bingo Card

36	24	18	30
21	45	28	32
22	9	12	42
49	16	25	8

Brain teaser: *What multiplication sentence uses the same number 3 times—as factors and as product?*

Ring-a-Round

You need: 2 markers, a number cube, and a coin

- You and your parent put your markers at START.
- Take turns rolling the number cube. Move your marker that number of spaces.
- Flip the coin. If it is HEADS, multiply the number in your circle by 2. If it is TAILS, multiply the number by 5.
- If your answer is correct, stay there. If incorrect, move *back* 2 spaces.
- The first one to get to the center is the winner!

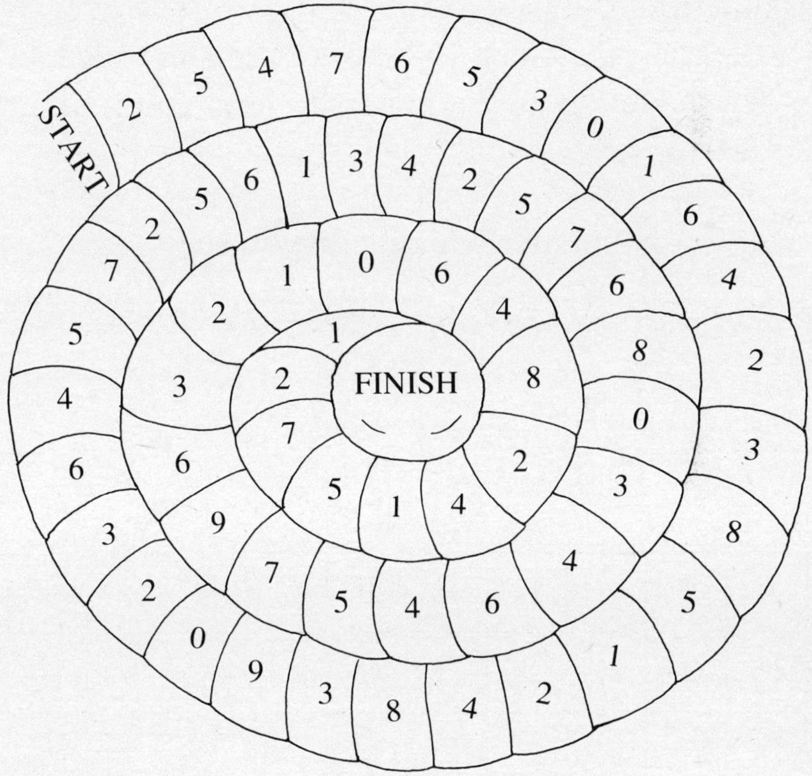

You can play this game again later using 3 and 4 as your factors.

The Riddle of Code Three!

Figure out the hidden meaning of this Code Three message!

- Find the products of the multiplication problems in the clues for this riddle.
- Then use the Code Breaker to translate each answer into a letter.
- Put the letters in the blanks to help you find out the Riddle of Code Three!

Riddle: No matter which side you see,
It takes all three to make me. What am I?

Answer: ___ ___ ___ ___ ___ ___ ___ ___

Clues:

$3 \times 7 =$ ______

$3 \times 4 =$ ______

$3 \times 8 =$ ______

$3 \times 5 =$ ______

$3 \times 3 =$ ______

$3 \times 9 =$ ______

$3 \times 2 =$ ______

$3 \times 6 =$ ______

CODE BREAKER

A	B	C	D	E	F	G	H	I	J	K	L	M
15	40	20	8	18	45	27	35	24	10	20	6	32
N	O	P	Q	R	S	T	U	V	W	X	Y	Z
9	17	4	30	12	35	21	36	25	50	22	16	14

Do Anything 4 Art!

You need: 9 colors of construction paper *or* plain paper and crayons, scissors, glue

1. You and your parent cut 8 strips of paper in each color. Make the strips about $\frac{1}{4}$ inch wide and 3 inches long.
2. Next, find these products.
   $4 \times 3 =$ ______ $4 \times 6 =$ ______ $4 \times 9 =$ ______
3. Ask your parent to help you divide the paper strips up into three groups. The number of strips in the groups should match the products you found.
4. Use the strips in each group to make a colorful pattern on plain paper. Be artistic!
5. When you have made a pattern that you like, glue it to the paper. You can keep it 4 yourself, use it 4 a gift to your parent, or put a 4 SALE sign on it!

Do this activity again! This time, cut out colored triangles. Make patterns for the products of 4×4, 4×5, and 4×7.

Example:

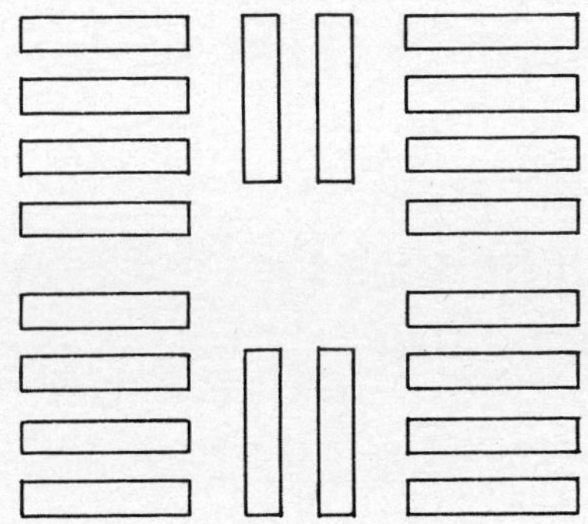

$4 \times 5 = 20$

Flip for It!

You need: 2 pencils, a coin, a number cube

You and your parent will FLIP over the picture you make by connecting the dots in this puzzle.

Take turns when you play.

- Roll the number cube and flip the coin to get the 2 factors you will multiply.
- If the coin comes up HEADS, multiply by 1. If it is TAILS, multiply by 0.
- What is your product? Draw that number of lines to connect the dots in this picture.
- The one who finishes the picture is the winner!

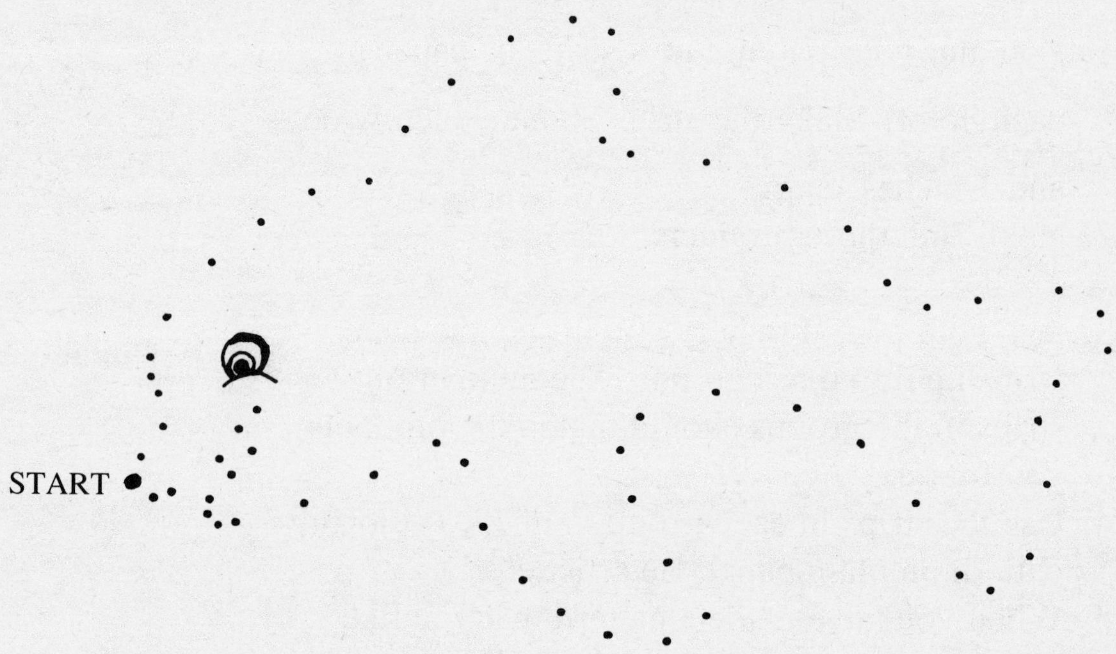

A puzzle: Zero is worth nothing, except when it is behind another number! Why?

*Slap*Factors!

For this game, ace = 1 and 10 = 0.

Shuffle the cards well, then sort them equally between you and your parent.

Each of you put your cards in a stack face down in front of you.

Then to play:

- At the same moment, both you and your parent turn a card over and SLAP it down in the middle of the table.
- The first one to say the product of the two factor cards gets to keep both cards. Put them in a separate stack.
- The one with the most cards at the end of the game is SLAP CHAMP!

Play again! This time remove only the jacks, queens, and kings. Then divide the cards so one person has aces through 5's and the other person has 6's through 9's.

You need: a deck of cards with the 7's, 8's, 9's, jacks, queens, and kings removed

A Farm Dinner

You need: a number cube, paper and pencil

Help the farmer decide how much food to buy the animals.

- First find the number of each kind of animal by throwing the number cube. Write the number rolled in the space after each animal's name.
- Then, figure out how much of each kind of food the farmer must buy. Identify the factors you multiply to solve each problem. Ask your parent to check every solution as you go along.

Animal Food

Kind of Animal	*Number of Animals*	*Amount Each Animal Needs*	*How much to Buy*
horses		5 pounds	
cows		5 pounds	
goats		4 pounds	
pigs		3 pounds	
sheep		4 pounds	
chicken		2 pounds	
ducks		2 pounds	
cats		1 pound	
dogs		3 pounds	

Snappy 6

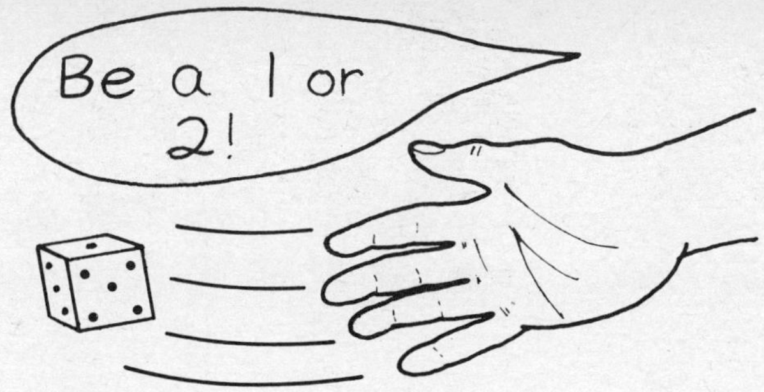

You need: a number cube

Take turns with your parent rolling the number cube.

Place the numbers you roll under the 6's at the right on this page.

After you have filled up all the spaces, ask your parent to time you to see how fast you can write the answers to these problems.

Try to finish your answers in less than one minute.

$6 \times \_\_$	$6 \times \_\_$	$6 \times \_\_$	$6 \times \_\_$
$6 \times \_\_$	$6 \times \_\_$	$6 \times \_\_$	$6 \times \_\_$
$6 \times \_\_$	$6 \times \_\_$	$6 \times \_\_$	$6 \times \_\_$
$6 \times \_\_$	$6 \times \_\_$	$6 \times \_\_$	$6 \times \_\_$

How much higher than 6 × 6 can you go?

A Maze of 7's

You need: 2 objects to be used as markers (beans, buttons, etc...)

Play for 5 to 10 minutes. Take turns dropping the 2 beans onto the board.

Give the 7's facts for both numbers the beans land on. For 49 and 14, you would say 7 × 7 and 7 × 2. Score 2 points for a correct answer.

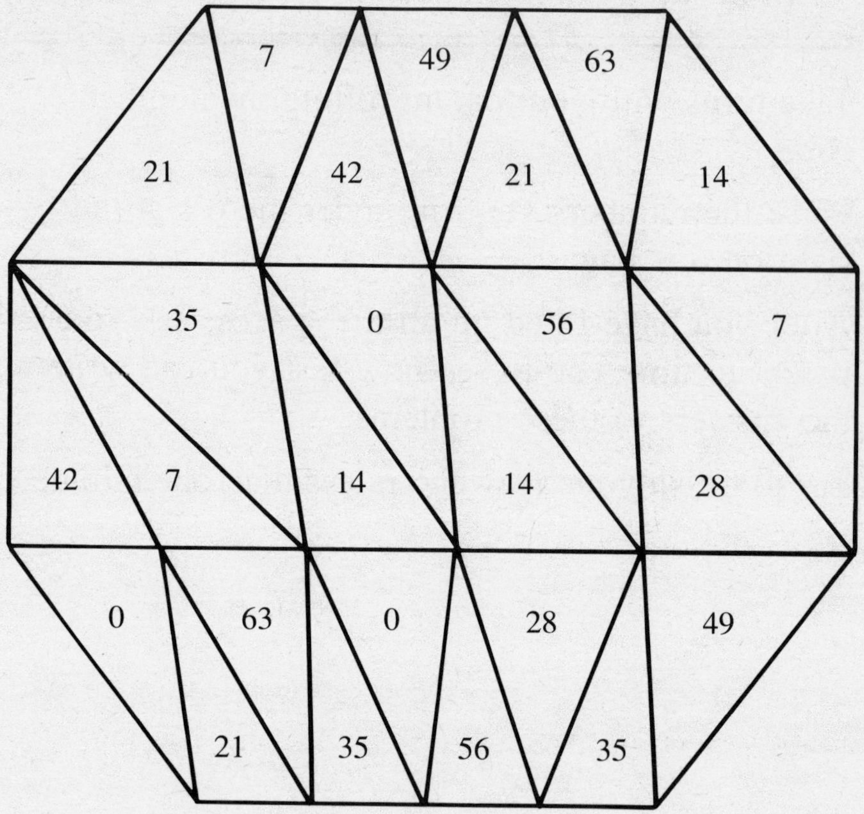

Choose a Number

You need: numbers 1 through 9 from the cutouts in the back of the book

Turn the cards face down.

Take turns playing this one with your parent. At each turn:

- Choose two numbers from the cut-outs to fill in the blanks in the word problems.
- Solve the problem.
- Do all 5 problems. Check each other's answers.

Do you solve problems the same way?

Talk about the ways you think.

Marian had ____ boxes of carrots. There were ____ carrots in each box. How many carrots were there?

Gino had ____ packs of balloons. There were ____ balloons in each pack. Paige has 40 balloons. Who has more?

Joe had ____ bags with ____ sticks in each bag. Sal had 27 sticks. Who had more sticks?

Kevin had ____ frogs in each bag. He had ____ bags. Jenny had 15 frogs. Who had more frogs?

Suzanne had a book. She reads ____ pages every ____ minutes. She read for about 30 minutes. How many pages did she read?

The Great 8!

You need: a pencil and a paper clip

Make a spinner by placing the pencil point in the middle of the 8 to hold the paper clip while it spins.

Then, to play:

- Spin the paper clip.
- Multiply the number you land on by 8 to find the product.
- Write the product in the space on the outer circle.

Keep playing until the wheel is completely filled in.

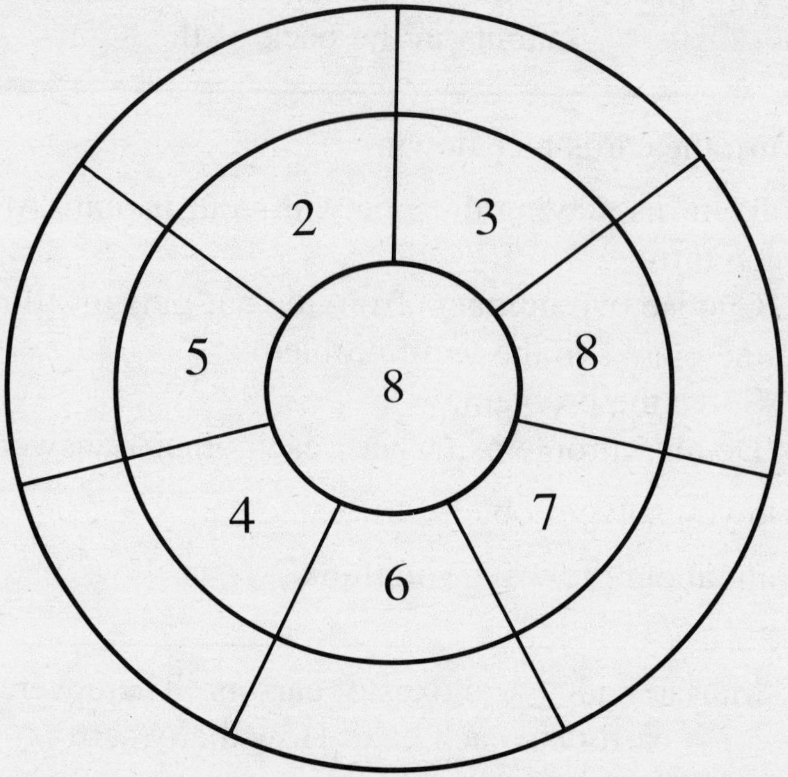

Was it hard to make the spinner stop on the last number?

FOUR WINS

- On the little slips, write the 9's multiplication facts (without the answers, as shown in the illustration above). Turn them face down.
- Choose one of the boards below.
- Take turns turning up a slip.
- Cover the answer of the multiplication fact if it is on your board. If it is on your board twice, you may cover the answer in both places.
- Turn the slip face down again and mix the slips up.

The first person to cover 4 in a row across, down, or diagonally wins the game.

You need: about 20 small objects to cover the spaces on these game cards, 10 same-size little slips of paper

9	81	27	36
54	0	54	45
63	9	72	18
27	45	36	18

Player 1

18	36	9	72
81	27	45	9
0	45	63	36
18	54	0	27

Player 2

Find the Way to Win

You need: a watch with a second hand

While your parent times you, see how long it takes you to go through the maze and multiply each number you see by 10. (Just say the answers.) Go as fast as you can. Catch your breath and go through a second time. See if you can improve your time.

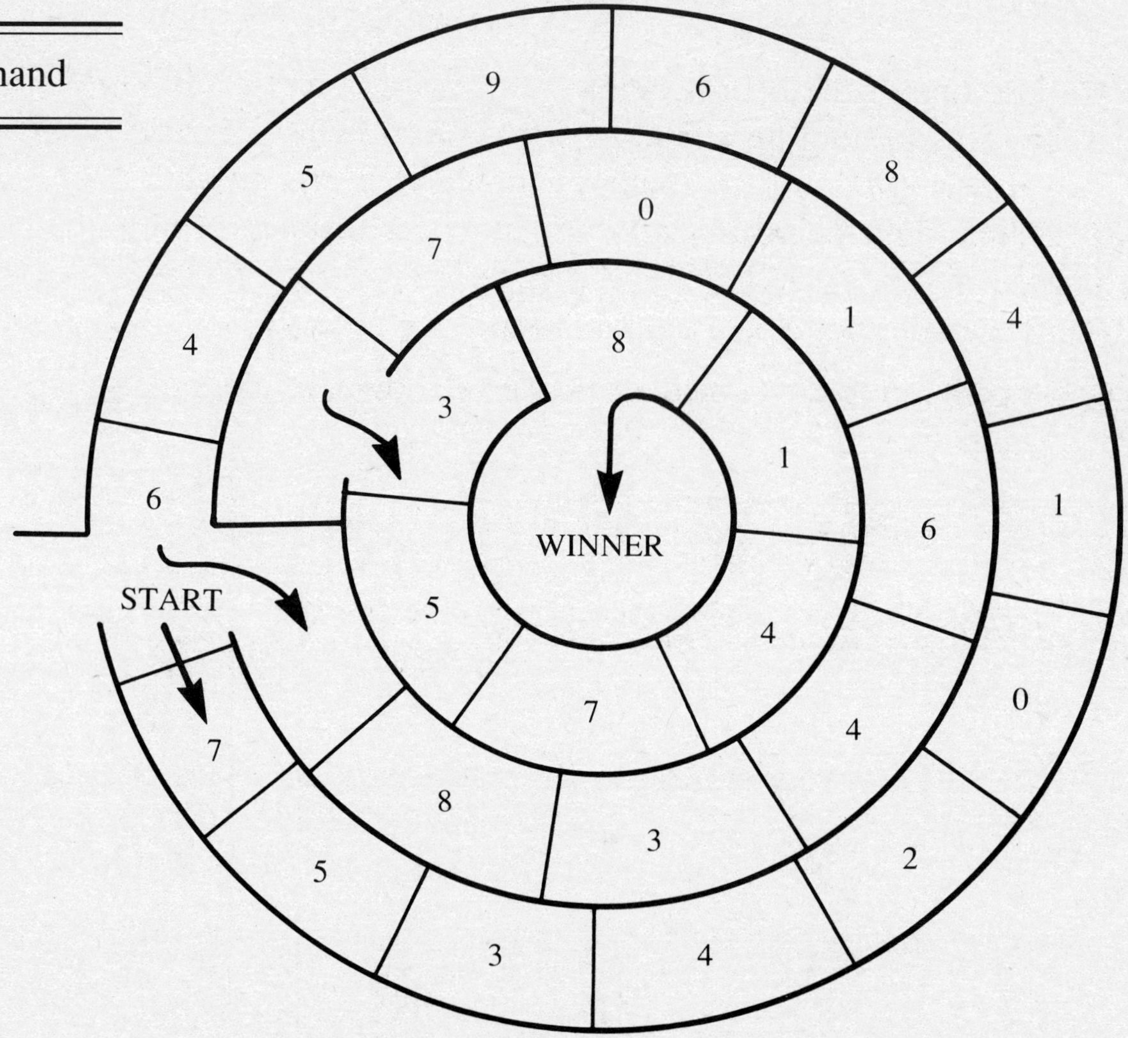

If there were a 10 on the maze, what would the answer be?

NUMBER STRATEGY

Take turns with your parent filling in the answers in the missing factors grids. Try to block the other person from getting all the answers in a whole grid. If one person writes a wrong answer, you may mark over his or her answer with your color ink.

You need: two pens with different ink colors

6's

24	=	6	×	
42	=	6	×	
30	=	6	×	

7's

35	=	7	×	
49	=	7	×	
14	=	7	×	

8's

24	=	8	×	
16	=	8	×	
64	=	8	×	

9's

81	=	9	×	
27	=	9	×	
45	=	9	×	

CARS AND TRUCKS

Extended Activity

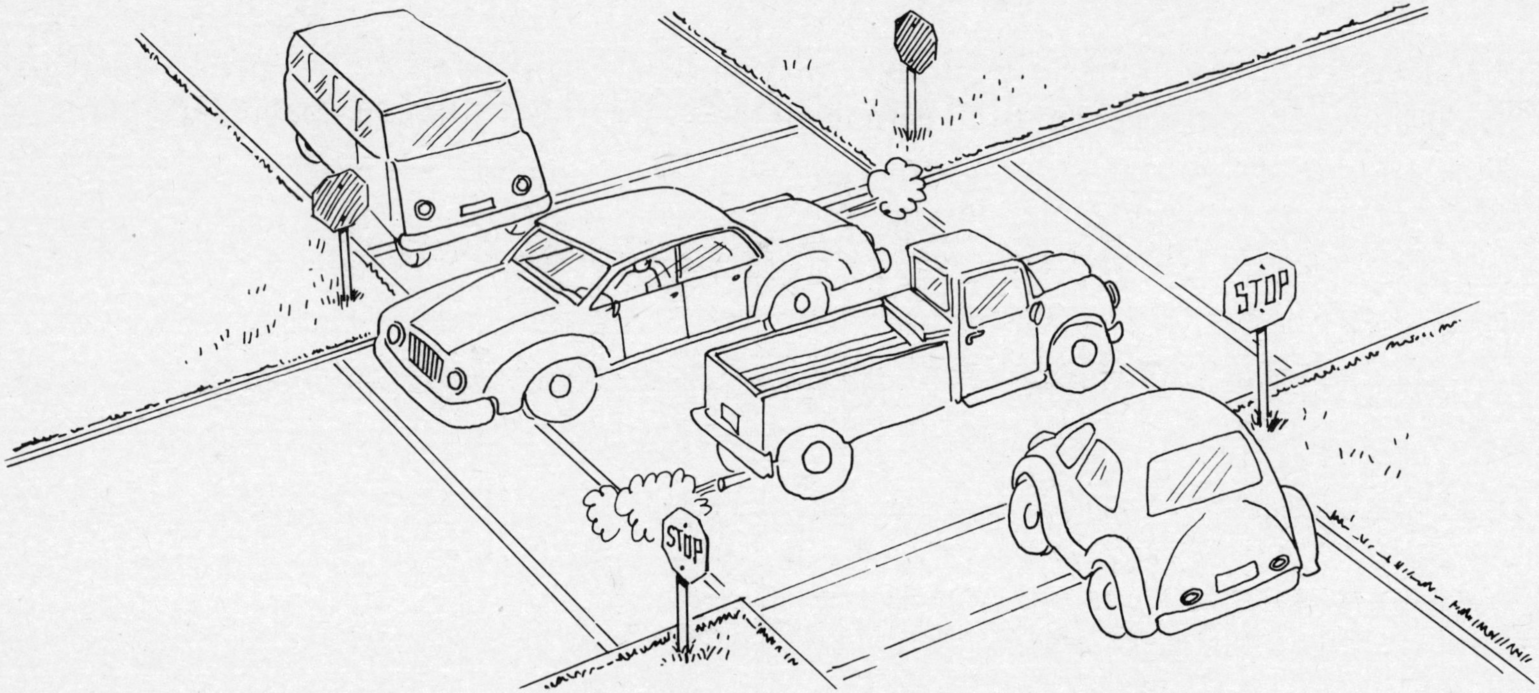

Cars and trucks and vans are built to carry different numbers of people.

Look at the pictures above and talk to your parent about the number of people each kind of vehicle was made to carry.

Take a 15-minute walk with your parent. While you walk, estimate how many people each vehicle you see can carry.

Record your findings on the special chart on the next page.

TALLY HO! (Cars and Trucks, page 2)

Before you go, fill out the second column of the table below. (Don't include busses—they'll make your work too hard.)

Type of Vehicle	*Estimated Number of People*	*Number of This Type of Vehicle You Saw*
sedan		
sports car		
van		
small truck		
semi		
other		

On your walk, then, work with your parent to keep a tally of each type of vehicle. Don't worry about how many people are actually *in* any vehicle. Don't think about how many *total people could be in all the vehicles*—that's for the next page, after you return home.

TOTAL HO! (Cars and Trucks, page 3)

Fill in columns 2 and 3 of the table below, after you get back. Then imagine all the seats were taken in each vehicle.

Use your multiplication facts to find how many people could have been carried by the different-sized vehicles. Ask your parent to check your answers.

Type of Vehicle	*Number of People*	*Number of Vehicles You Saw*	*Total Number of People Who Could Be Carried*
sedan			
sports car			
van			
small truck			
semi			
other			
		Grand Total	

Toothpick Time

You need: 30 or more toothpicks or small strips of paper

Use your toothpicks for counters. Let each toothpick equal 1.

Call out division problems to your parent and let him or her show the problem using the counters. You tell the answer by counting the groups.

Then have your parent call out problems to you. This time you show the problem using the toothpick counters. Your parent will tell the answers.

Here are two to get you started.

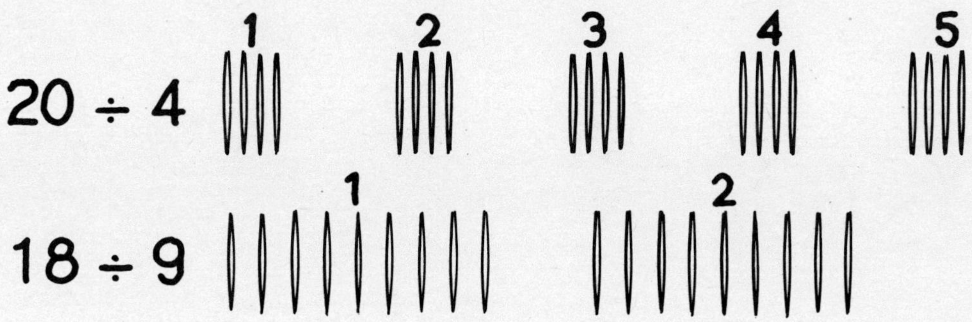

Can you show 1 ÷ 1? Try it!

Division Fill In

You need: two colors of markers

Take turns with your parent filling in the division mazes.

Begin at START, and fill in the chains answer by answer. Check each others' answers.

If you miss one, your parent has a chance to fill in the answer with her or his color.

The person with more answers in her or his color at the end wins the game.

START HERE
Player 2

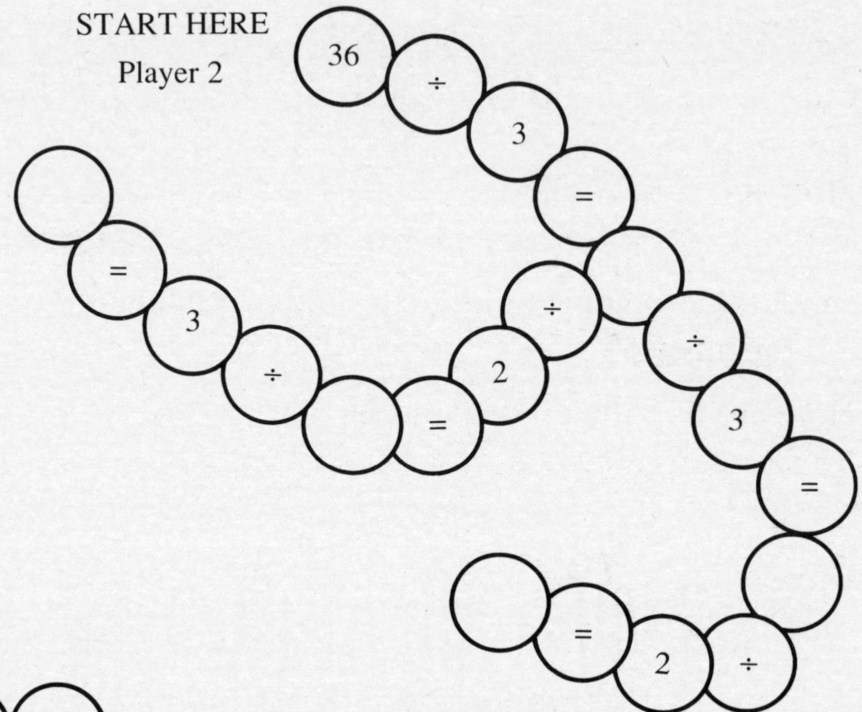

START HERE
Player 1

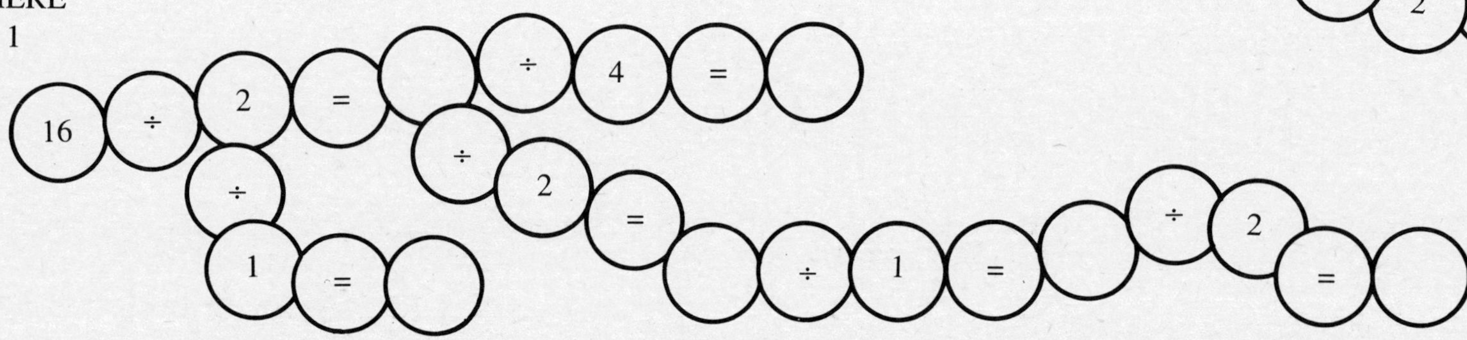

How Fast Can You Think?

You need: paper, pen, scissors, a watch

Cut the paper into small squares, about the size of the 6- and 7-squares shown.

Write each of the numbers at the right on separate squares of paper.

Mix and place the numbers face down.

Then have your parent time you as you play:

- Pick up the numbers one at a time.
- As you pick up a number, decide if it is divisible by 6 or by 7. Place it in the correct stack on this page.
- When you have finished, your parent will check your answers and announce your time.

Play again later and see if you can beat your time.

Numbers to put on cards:

6, 7, 12, 18, 24, 28, 30, 35, 36, 42, 42, 48, 49, 54, 56, 60, 63, 70

6	7

Into which group did you put 42? Why?

4 4 4 4 in a Row

Drop a bean onto the playing board. Make a division problem using the number you landed on and either an 8 or 9.

For example, if your bean lands on 32, you say, "32 divided by 8 is 4." If you are correct, put your initials on that square. The first player to initial 4 squares in a row is the winner.

Try not to land on a space that has already been used.

If you do, you lose a turn.

You need: 1 bean per player, pencil

32	18	56	16
45	81	54	36
48	63	40	64
64	27	24	72

Which is the only number that you could have used with either 8 or 9?

Pyramid Popcorn

You need: several popcorn kernels

Select the game board you like. Your parent takes the other.

Take turns working your way up your own pyramid, problem by problem.

Try to solve your problems correctly but as quickly as possible.

As you say a problem correctly, place a popcorn kernel on it. If you get stuck on one, skip it and go back to it at the end of your game.

If your parent permits, pop some corn to eat for being such a good divider.

$24 \div 6 =$

$36 \div 6 =$ | $21 \div 7 =$

$35 \div 7 =$ | $14 \div 7 =$ | $6 \div 6 =$ | $28 \div 7 =$

$12 \div 6 =$ | $42 \div 7 =$ | $49 \div 7 =$ | $42 \div 6 =$

$7 \div 7 =$ | $18 \div 6 =$ | $30 \div 6 =$

$6 \div 6 =$

$42 \div 6 =$ | $28 \div 7 =$

$7 \div 7 =$ | $36 \div 6 =$ | $12 \div 6 =$ | $21 \div 7 =$

$35 \div 7 =$ | $18 \div 6 =$ | $42 \div 7 =$ | $24 \div 6 =$

$30 \div 6 =$ | $14 \div 7 =$ | $49 \div 7 =$

Get the Apples

You need: 2 buttons for markers, a number cube

Take turns with your parent as you play this game.

- Roll the number cube and move up the ladder on the left that many spaces.
- Solve the problem. (If you land on a rung that is not a problem, follow those directions.)
- Try to move up one ladder and down the other ladder without making a mistake.

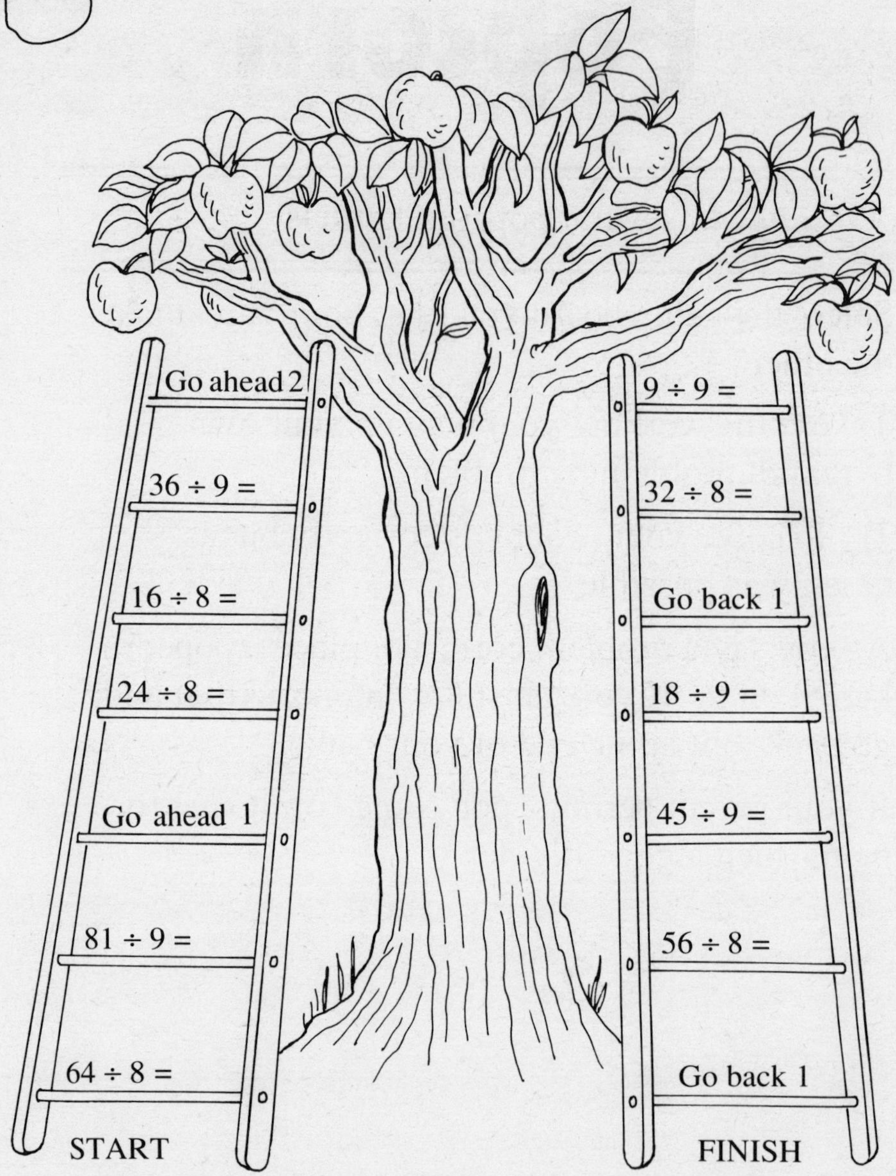

Fence the Schoolyard

This page is your schoolyard. It's getting a new fence. One side has already been done, but you need to finish the other three sides.

- Show a different division fact on each side.
- Your parent can help by drawing fenceposts for you.
- You darken some of the posts to show how you divided them into groups.

Draw your house and fence your yard with different division facts.

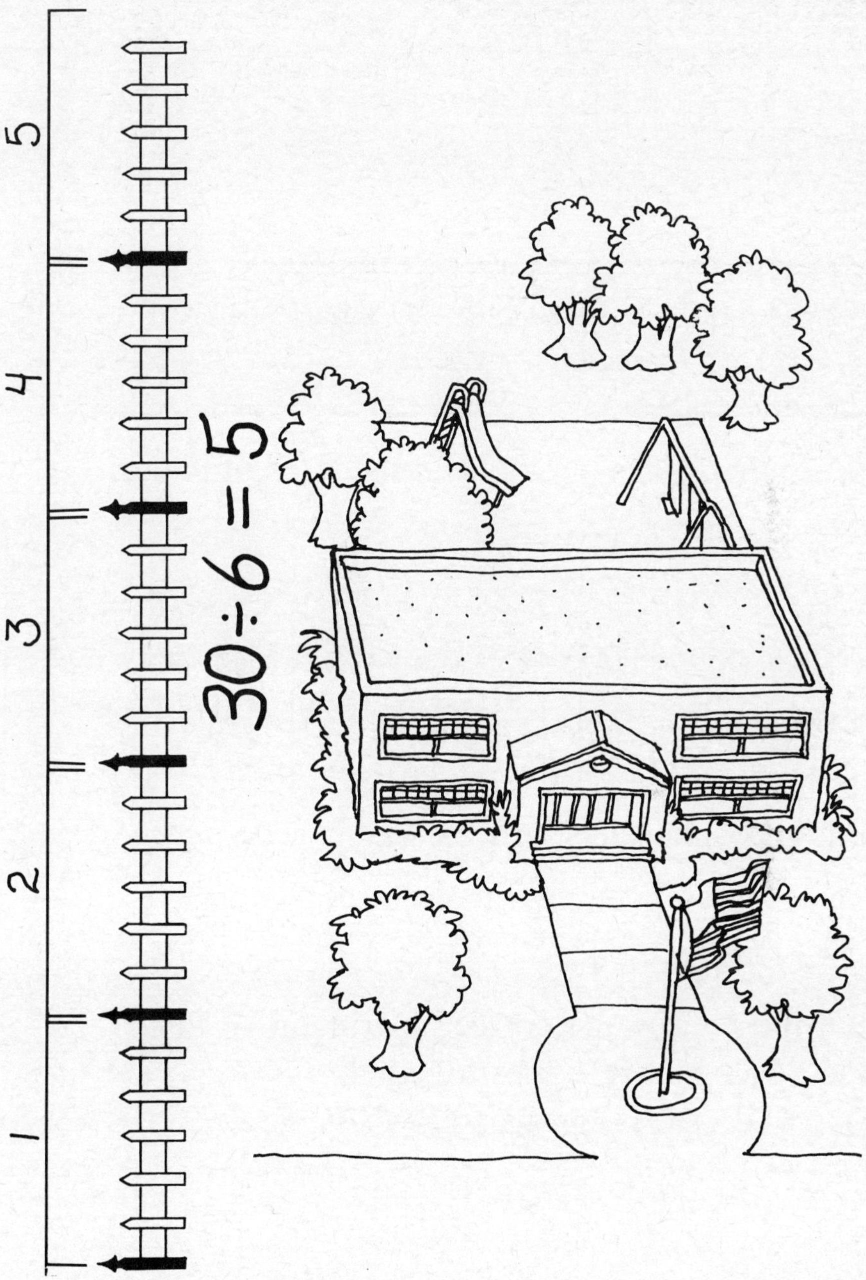

Put the Families Together

You need: a small bag, 20 small slips of paper

Write the following facts on the slips of paper.

$4 \times 8 = \square$	$\square \div 5 = 3$	$4 \times 6 = \square$	$\square \div 4 = 8$
$\square \div 3 = 6$	$8 \times 4 = \square$	$2 \times 5 = \square$	$\square \div 4 = 6$
$5 \times 3 = \square$	$6 \times 3 = \square$	$\square \div 6 = 3$	$5 \times 2 = \square$
$\square \div 5 = 2$	$\square \div 6 = 4$	$\square \div 3 = 5$	$3 \times 6 = \square$
$6 \times 4 = \square$	$\square \div 2 = 5$	$\square \div 8 = 4$	$3 \times 5 = \square$

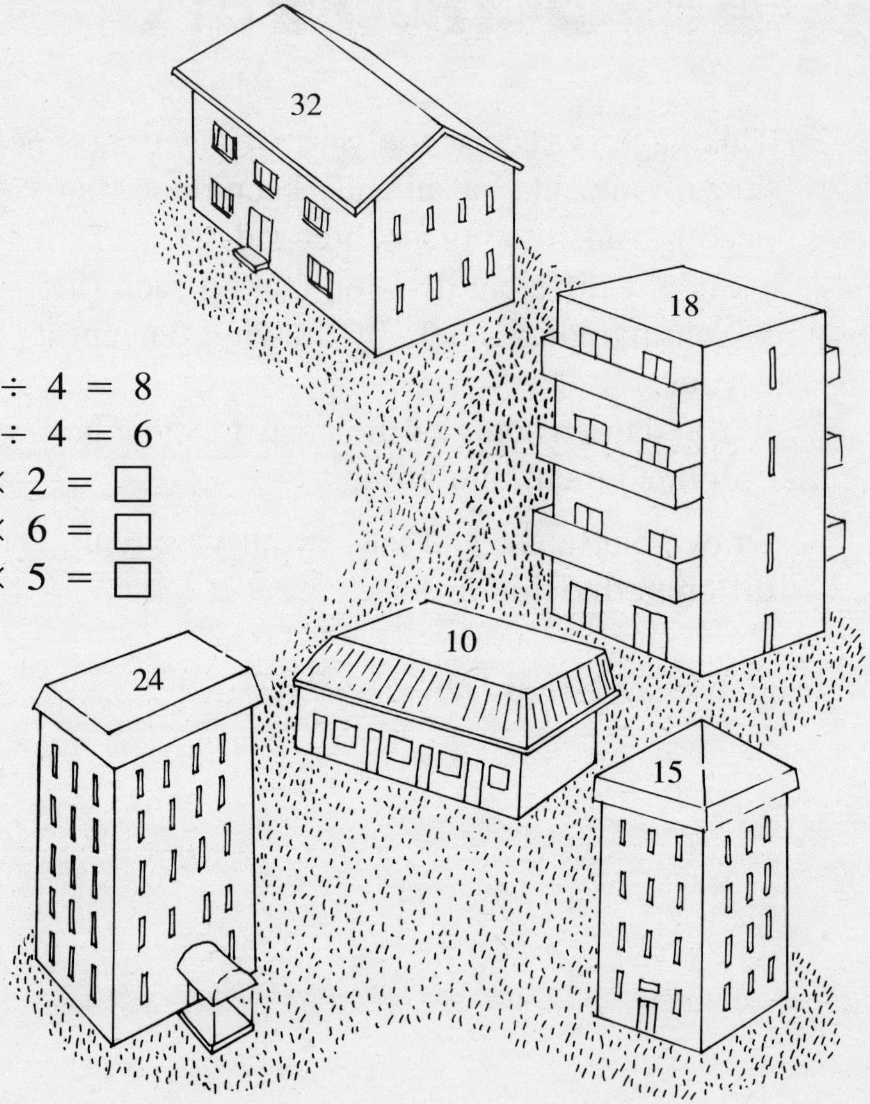

Put all of the facts into the bag.

Take turns choosing a fact out of the bag.

Solve the problem.

Put the fact on the correct apartment so that the "fact family" goes together.

Example of a "fact family"

$5 \times 6 = \boxed{30}$ $\quad \boxed{30} \div 5 = 6$
$6 \times 5 = \boxed{30}$ $\quad \boxed{30} \div 6 = 5$

WHEEL DIVISION

You need: a paper clip and pencil for each wheel

You and your parent both place a pencil in the middle of one wheel to hold a paper clip in place.

- Spin both paper clips.
- Take turns dividing the greater number by the lesser number.
- Keep a running tally of the answers that have remainders.
- The first person to reach 20 with his or her remainders wins the game.

My remainders:	
Your remainders:	

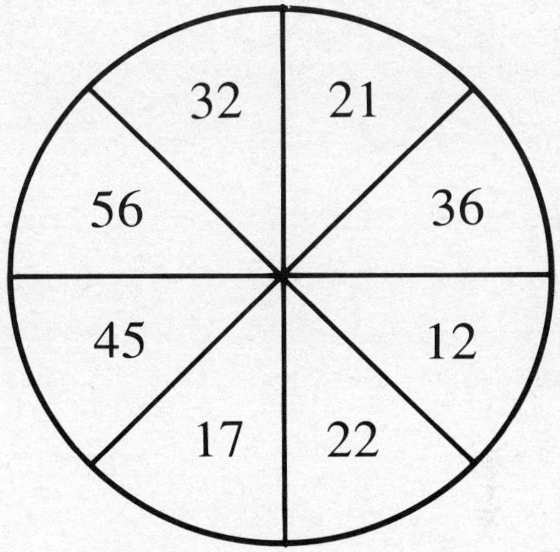

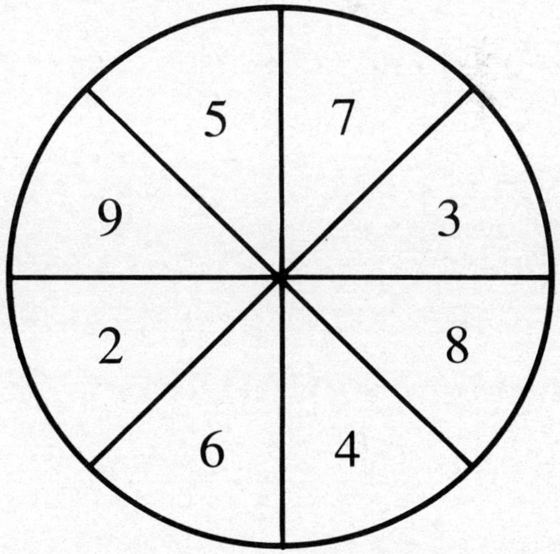

How Many Objects Are There to Go Around?

Go house exploring with your parent. Find things in your home that everyone in your family uses. Here are the things you will be looking for:

glasses or cups	towels	shoes
plates	toothbrushes	bathrooms
closets	spoons	rooms

Use the chart on the next page to keep a record of your findings.

Count the number of each of the objects. Then divide that number by the number of people in your family. Your parent may round some numbers so that the problems will divide evenly.

Then sum up your discoveries on the third page.

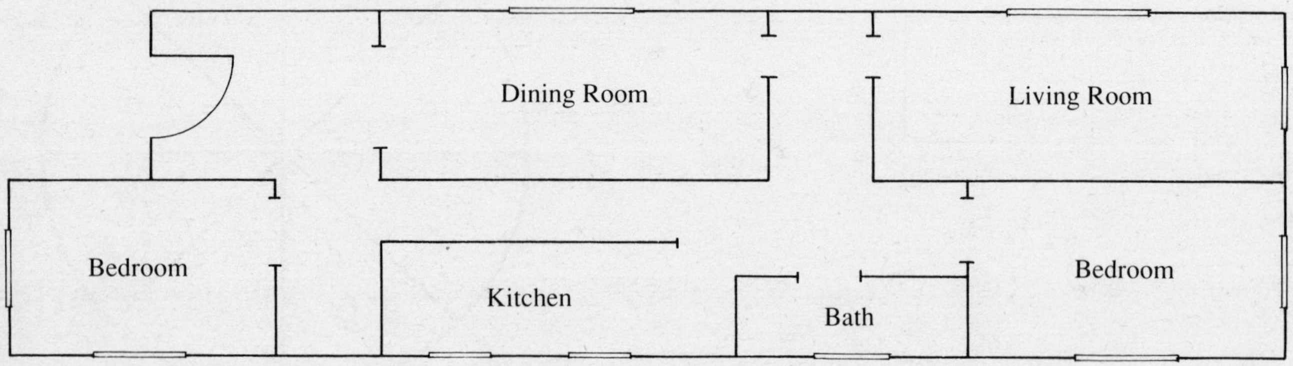

(How Many Objects..., page 2)

MY COUNTING CHART

Objects	*Number of Objects*		*Number of People*		*Number per Person*
glasses		÷		=	
towels		÷		=	
shoes		÷		=	
plates		÷		=	
bathrooms		÷		=	
toothbrushes		÷		=	
closets		÷		=	
spoons		÷		=	
rooms		÷		=	

MY SUMMARY

Tell what you learned after exploring your house. Did you find that there were a lot of objects for each person, or did you discover that each person had just the right amount of objects? Write a short paragraph about your house exploration.

Part 4

Fractions and Decimals: is it one-half full or 0.5 empty?

A note to parents

Your child knows the difference between a full cup of milk and one that's only half or one-third full. That one-half or one-third could also be measured in decimals, in the same way that parts of a gallon of gasoline are computed.

Figuring fractions and decimals sometimes can be frustrating when it's just moving numbers around on paper. Help your child see that in real-life situations, these skills can be fun!

You can try the following ideas to show your child how common it is to use fractions and decimals in everyday life.

- When at the grocery store, show your child the decimals used in marking quantities and prices for food items. Have him or her read the decimals to you.
- If you are measuring food for cooking, let your child tell you whether the cup is one-third or one-fourth full.
- Practice dividing whole things into sections (fractions), such as an orange into segments or bread into slices.

Now use these activities to help your child on the way to a knowledge of how to work with fractions and decimals.

Blooming Fractions

Ask your parent to help you, as you follow the directions below to make a colorful fraction garden.

- Color $\frac{1}{2}$ of the flowers red.
- Color $\frac{1}{4}$ of the flowers yellow.
- Color $\frac{1}{8}$ of the flowers pink.
- Color $\frac{1}{8}$ of the flowers orange.
- Color $\frac{10}{16}$ of the leaves green.
- Color $\frac{6}{16}$ of the leaves brown.

You need: red, yellow, orange, brown, pink, and green crayons, markers, or colored pencils

Count the pink flowers. How many are there? ______

How many yellow flowers are there? ______

How many red flowers are there? ______

How many brown leaves are there? ______

Mobilizing Fractions

You need: construction paper—plain or colored; 3 yards of string or yarn; a ruler; markers, crayons or colored pencils; scissors; clothes hanger

1. Ask your parent to help you cut out three each of circles, triangles, and squares. They should be approximately 4 inches across.
2. Color in the regions listed below on one of each of the shapes you have cut out. Color the region on both sides.
   - color $\frac{1}{4}$ of the shape
   - color $\frac{1}{2}$ of the shape
   - color $\frac{1}{3}$ of the shape
3. Punch out a hole at the top of each shape.
4. Cut the string into three 1-yard lengths.
5. Loop the string through the hole.
6. Slide down a foot on the string and knot it through the hole of the next shape.
7. Repeat Steps 5 & 6 for the other shapes.
8. Attach the strings to a hanger.

You have just made a fraction mobile!

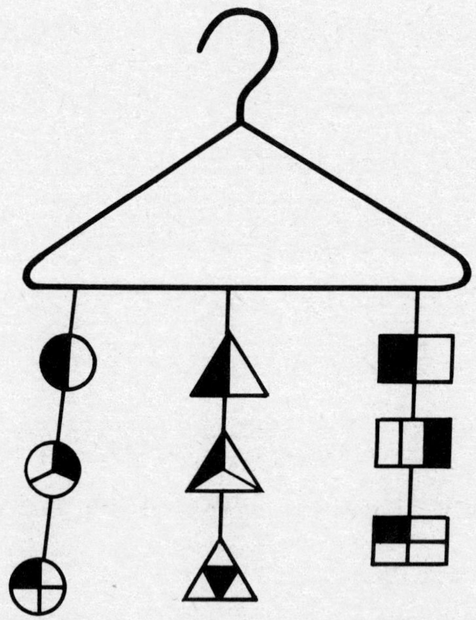

Baggin' It

You need: a paper grocery bag; crayons, markers, or colored pencils; scissors; a ruler

Instructions:

- Cut out the two big flat sides of the bag while your parent watches.
- Lay one side down on a flat surface.
- With your parent's help, mark one side in thirds.
- Color one third of that side.
- Get out the other side.
- Mark this side in fourths.
- Color one fourth of that side.
- Lay the two sides you marked side by side.
- Compare the colored areas. Which is greater, $\frac{1}{3}$ or $\frac{1}{4}$?
- Color in another $\frac{1}{3}$ and another $\frac{1}{4}$ section.

  Which is greater now, $\frac{2}{3}$ or $\frac{2}{4}$?

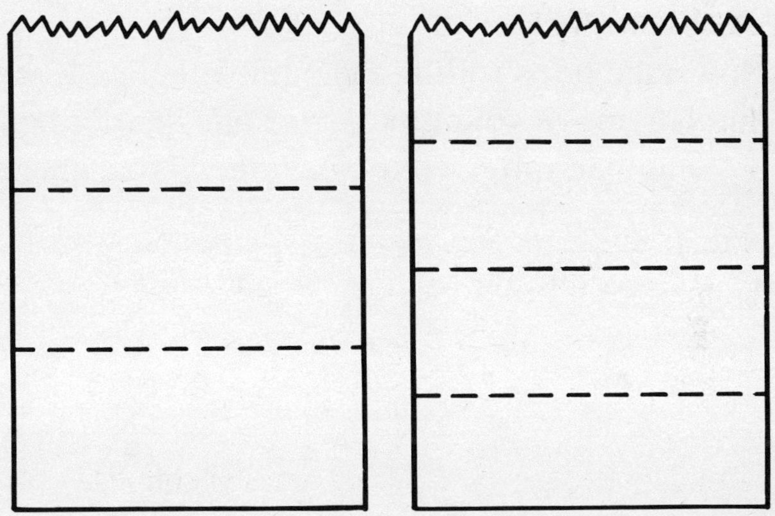

Hickory Dickory

This is a game about quarter hours and using the clock. You will have a good time playing this game with your parent. Just follow the simple rules!

- Each player uses one of the clock faces for the game. Each person needs one short and one long toothpick to use as the clock hands.
- Place the clock hands at 12:00.
- Now, take turns rolling the number cube. If you get an even number, move your clock *back* one quarter hour. If you get an odd number, move your clock *ahead* one quarter hour. The first player to reach 3:00 wins!

You need: a number cube, four toothpicks (2 short and 2 long), clocks on this page

Each time you move your clock ahead or back, say "The clock struck _____." (Say the time)

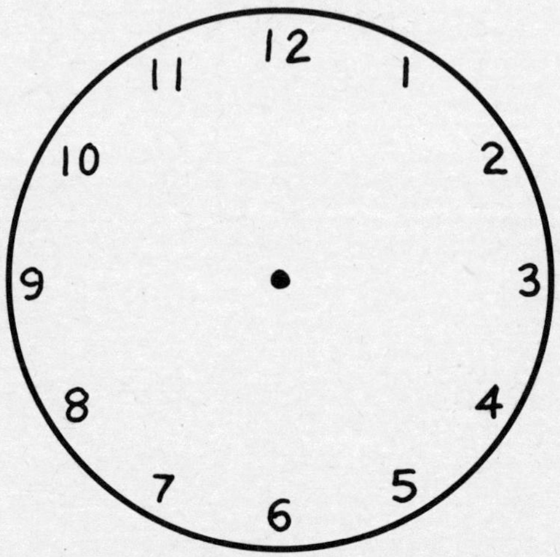

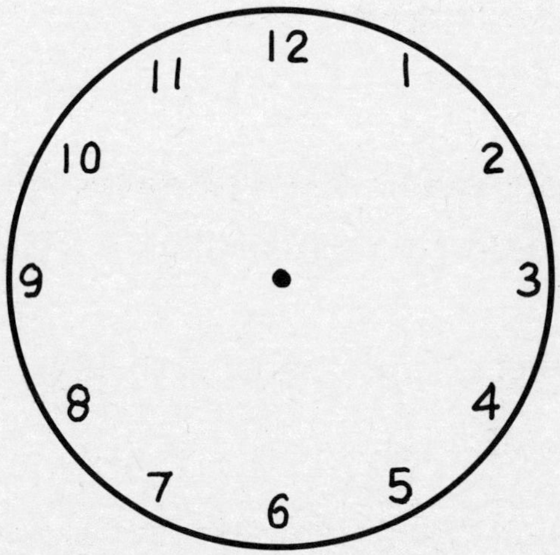

We use fractions everywhere. The gauge in a car or any other vehicle uses fractions to show the amount of fuel in the tank. Look at the gauges below. Use a fraction to tell how much fuel is in each tank.

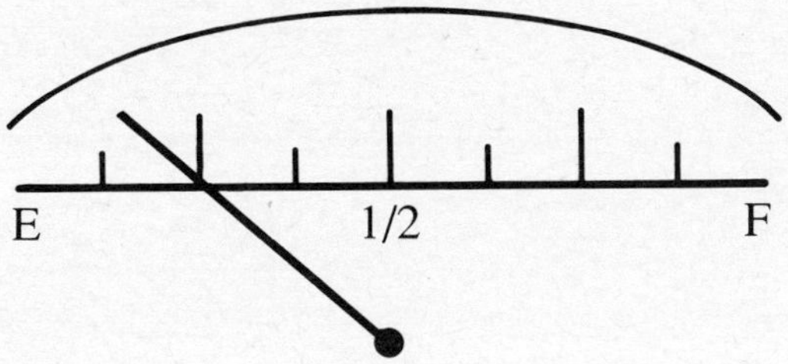

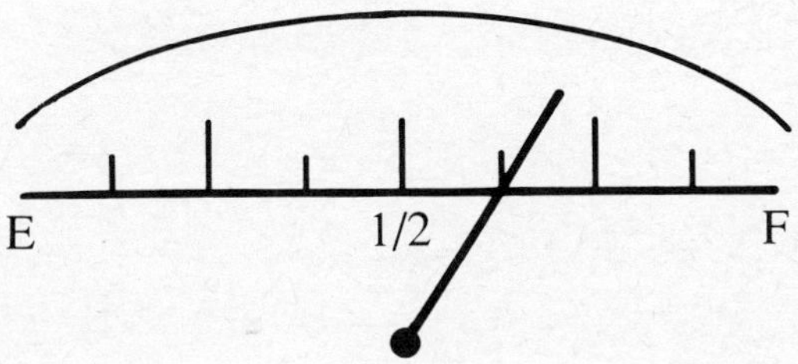

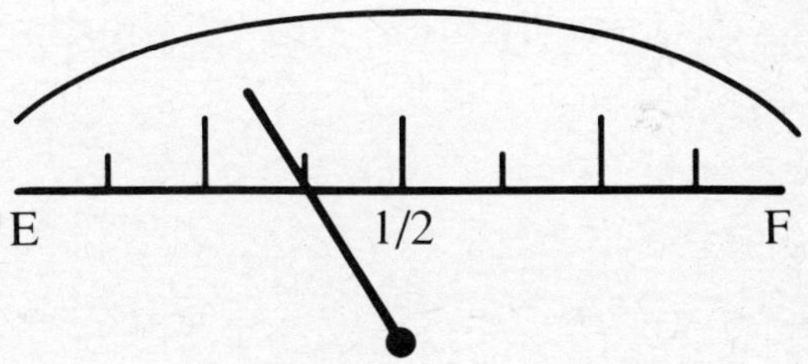

________ ________ ________

Do you have a car or do you know someone who does? Do you ride the bus? Ask your parent to help you check the fuel gauge on a vehicle. Mark the amount of fuel you saw on the gauge below.

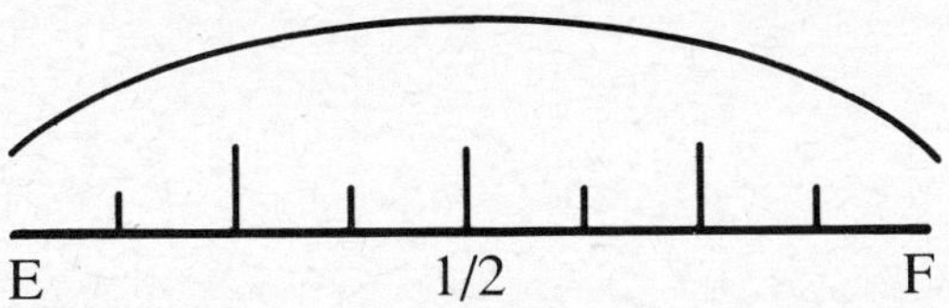

Hint: Remember

$$\frac{1}{2} = \frac{2}{4} = \frac{4}{8}$$

$$\frac{3}{4} = \frac{6}{8}$$

Tasting Fractions and Whole Banana Numbers

Extended Activity

Put fractions to work with your parent by making the following recipe.

These are the things you are going to need to gather together before you begin your tasteful project. Using fractions in the kitchen can be fun and fattening, so keep your fingers out of the batter, and make this delicious recipe.

You need these tools:

- a large mixing bowl
- a medium mixing bowl
- a 1-cup measuring cup (marked in fractions)
- measuring spoons
- 2 loaf pans
- a mixer or rotary beater
- a large spoon or spatula

You need these ingredients:

$1\frac{1}{2}$ cups sugar

$\frac{3}{4}$ cup shortening, margarine, or butter

$4\frac{1}{2}$ ripe bananas

3 eggs

$1\frac{3}{4}$ cups flour

$\frac{3}{4}$ teaspoon salt

$2\frac{1}{2}$ teaspoons baking powder

MIX, MASH, AND MEASURE (Tasting Fractions . . . , page 2)

Ask your parent to turn on the oven and set it at 350 degrees. Wash your hands and . . .

Mix

- Measure $1\frac{1}{2}$ cups sugar into the large bowl.
- Add $\frac{3}{4}$ cup shortening or softened butter or margarine.
- Cream the sugar and shortening until fluffy. (Ask your parent to help with this part. Creaming can be done with the mixer or with a rotary beater, or even with two forks.)

Mash

- Mash $4\frac{1}{2}$ ripe bananas.
- Beat 3 eggs.
- Put the bananas and eggs together and mix with the creamed mixture.

Measure

- Measure $1\frac{3}{4}$ cups flour in the medium bowl.
- Add $\frac{3}{4}$ teaspoon salt.
- Add $2\frac{1}{2}$ teaspoons baking powder.
- Sift the dry ingredients together. (Ask a parent to help again.)

Mix Again

- Carefully add the dry ingredients to the egg and sugar mixture in the large bowl.
- Mix with a spoon until just blended.
- DON'T OVERMIX!

BAKE, WAIT, AND EAT (Tasting Fractions . . . , page 3)

Bake

- Grease the two loaf pans with shortening.
- Divide the batter in half, and put each half in a pan.
- Ask your parent to put the pans in the oven.

Wait

- Now you have to wait for about 35 minutes. While you are waiting, draw a picture of yourself and your parent working in the kitchen, learning fractions through cooking.

Eat

- After 35 minutes, ask your parent to check the banana bread by testing it in the middle with a toothpick. If the toothpick comes out clean, the bread is ready.
- Now you can eat your work! As the bread cools and you eat a slice, imagine how many slices you could cut from the two loaves. What fraction would one slice be?

Penny Pinching

Add these sums of money.

When you say the answer, say "hundredths" instead of cents. When you write the decimal, think of dollars and cents.

Add the money and write the decimal it represents:

2 dimes + 3 pennies = 0.23

6 pennies + 6 dimes = ______

5 dimes + 8 pennies = ______

3 dimes + 2 pennies = ______

4 pennies + 8 dimes = ______

1 dime + 0 pennies = ______

9 pennies + 1 dime = 0.19

4 dimes + 0 pennies = ______

9 dimes + 1 penny = ______

4 pennies + 7 dimes = ______

10 pennies + 0 dimes = ______

10 dimes + 9 pennies = ______

Use your money to make these numbers:

$\frac{23}{100}$ $\frac{58}{100}$ $\frac{84}{100}$

Say the amounts in hundredths, as you did above.

You need: 10 pennies and 10 dimes (you can use real coins or the cut-outs from the back of the book)

The Price Is Right

Decimals are used in the prices we see on price tags. The number before the decimal point is a whole number and the number after the point shows the dollar broken into hundredths, or cents.

Fill in the blanks below.

Ask your parent to help you read the price tags.

Say "___ whole dollars and ___ hundredths" or "___ dollars and ___ cents" for each tag you read.

Make price tags of your own. You and your parent pretend that you have a store. Price some of the items in your room. Make tags for the items using decimals.

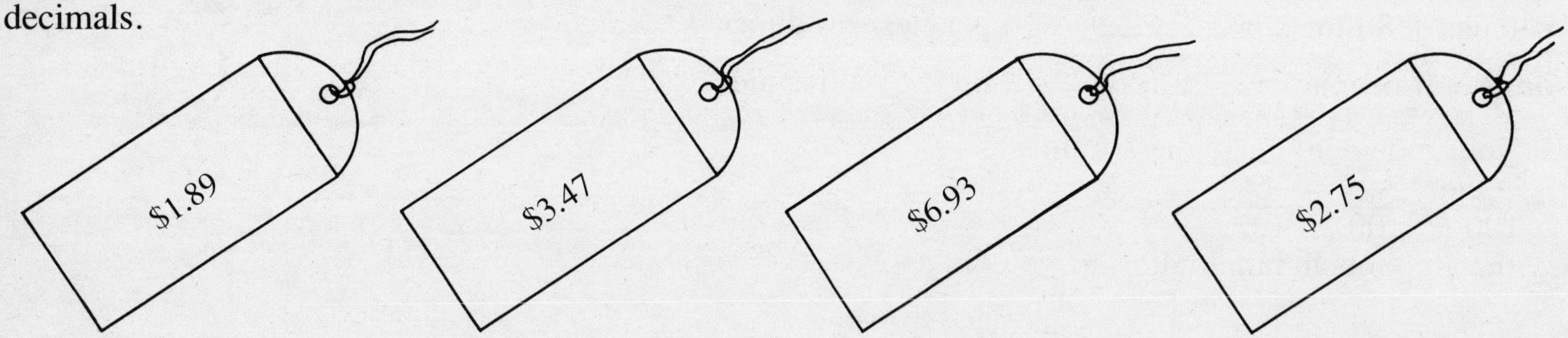

___ dollars and	___ dollars and	___ dollars and	___ dollars and
___ cents (hundredths)	___ cents (hundredths)	___ cents (hundredths)	___ cents (hundredths)

Turn Your Radio On!

You need: crayons or colored pencils

Billy's favorite radio stations are KVOE, 103.2; WOAW, 95.6, and KBIA, 100.4. Each station can be found on the radio dial below. The radio stations use whole numbers and decimals so the listeners can find the stations. With your parent's help, locate the stations Billy likes on the dial below. Put a different colored mark for each station number. See the example below.

103.2

95.6

100.4

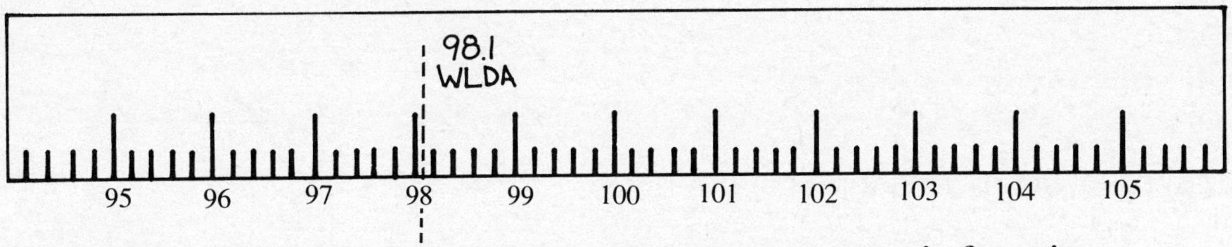

Fill in the radio dial below with the numbers of your and your parent's favorite stations. List the numbers you are using in the blanks. Label your stations like the one above.

____

____

____

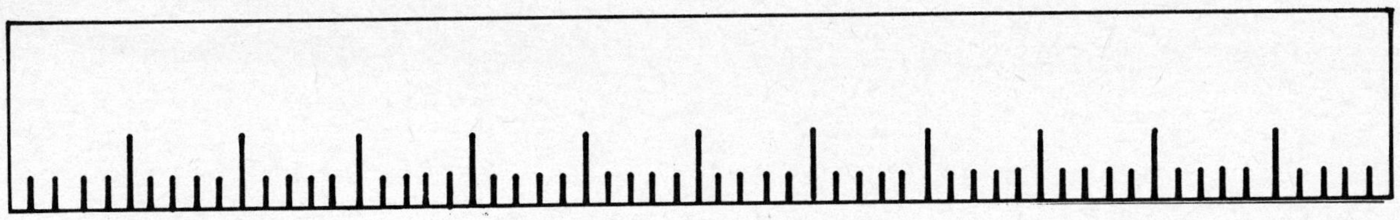

Bean Piles

Take turns as you play this game with your parent.

- Spread the beans on the table.
- Throw the number cube to find out how many piles of beans to make. (If you get a 1 or 2, throw again—that's too easy.)
- Divide the beans into equal piles. (You may have a remainder.)
- Let your parent tell you how many piles to count.
- Count the total number of beans in the piles you have been asked to count.
- Write a fraction showing the counted beans.
- Turn the fraction into a decimal.

You need: 100 dry beans or toothpicks, a number cube, paper and pencil

What food do you divide into equal groups when your family eats?

Measure Me

You need: inch ruler

Ask your parent to help you make these measurements.

The length of my hand: ___________

The length from my elbow to my wrist: ___________

The length of my nose: ___________

The length of my foot: ___________

The length from my knee to my heel: ___________

My height: ___________

You can use body measurements to estimate other measures.

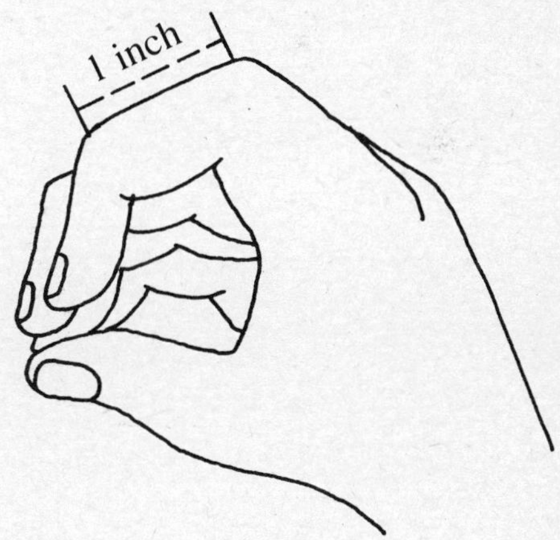

See if you or your parent match either of these approximate measures:

- From the tip of your nose to the tip of your middle finger, with arm outstretched, is about 1 yard.
- The middle section of your index finger, from joint to joint, is about 1 inch.

ALL AROUND THE HOUSE

Ask your parent to help you find the perimeters of some rooms in your house.

One of you measures.

The other writes down the measurements.

Show the measurements on the chart.

Room	*Length*		*Width*		*Perimeter*	

Perimeter = length + length + width + width

You need: a yardstick or measuring tape

If a room is square, you can find the perimeter by measuring only one side. Explain how.

Part 5

Probability and Geometry: pure chance, probable chance, and the "plane" truth

A note to parents

Have you ever noticed how easy it is to explain to a child how you predicted the winning team in the football playoffs? (Of course, it was the team that had won its last six games in a row.) Or how you "knew" that your lawnmower would never start again? (after it had failed to start the last eight times you tried to use it) These are two examples of the ways probability and chance work in our everyday lives. In Part 5, your child will learn that logical predictions can be made based on bits of information.

To help your child become more comfortable with the ideas of probability and chance, try these simple activities.

- Ask your child to predict who the next telephone call might be for at a certain time of day. She or he can use past experience to make a shrewd guess.
- When playing board games, encourage your child to try to predict possible outcomes. These predictions often will work out, but reassure your child if the predictions do not actually come true. Explain that chance is just that—chance.

You can help your child explore geometric shapes by doing these simple activities.

- Point out examples of geometric figures in everyday life. For example: Cereal boxes are rectangular solids; doors usually look like rectangles; windows, tables, and pot holders sometimes are square.
- Help your child find examples of angles around the house. Some common examples include the corner of a room, a partially opened door, the edge of a table, an open book.

Now try the activities in Part 5.

Fill It Up

Do this activity at your kitchen or bathroom sink. Ask your parent to help.

Choose two cups close to the same size. Tell which one you think will hold more water. Then fill it with water. Empty it into the other smaller one, and see if it runs over. Did you guess right?

Compare other containers the same way. Clean up when you finish.

You need: several cups or containers of different sizes

MEASURE FOR MEASURE

Here is a way to make your own measuring set.

- Fold the piece of paper in half, cross-wise. Ask your parent to cut it along the crease mark. Write GALLON on one part. Put the part marked GALLON aside.
- Fold the other piece of paper in half. Again ask your parent to cut along the crease. Write HALF-GALLON on one of these parts. Put the part marked HALF-GALLON aside.
- Fold the remaining piece cross-wise, in half. Have this piece cut also. Write QUART on one of the parts, and put it aside.
- Once again fold the remaining piece cross-wise, in half. Again your parent cuts, and again you write on one part—the word PINT. Put that part aside.
- Finally, fold the clean piece one last time and cut. This time write CUP on one of the pieces. Discard the other piece.

You need: a piece of paper about the size of this page, scissors, a pencil

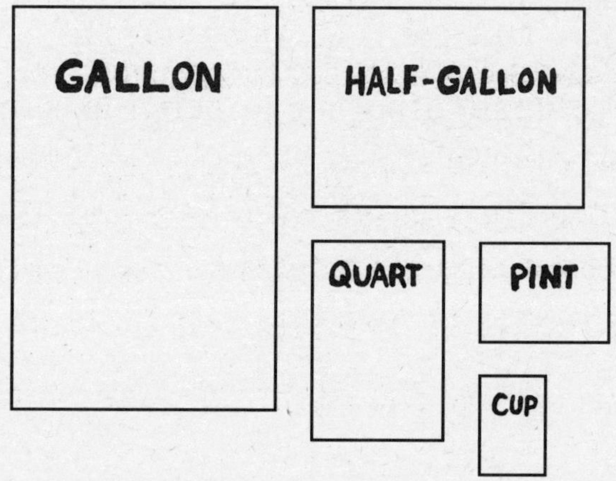

Now ask each other questions like "How many pints in a quart?" "How many cups in a pint?" See how many times the smaller measure fits on the larger.

Light or Heavy?

You need: a paper clip, a liter of water in a bottle (A large soda bottle holds 2 liters. Fill one half full.)

We know that a paper clip weighs about one gram and that a liter of water weighs about one kilogram.

Explore your home and find 5 things that weigh about one gram. Find 5 things that weigh about one kilogram. (Use a paper clip and a liter bottle of water to compare if you like.)

Ask your parent to check your choices. Does he or she agree with your guesses?

Ask your parent to find 2 objects that weigh about one gram or one kilogram.

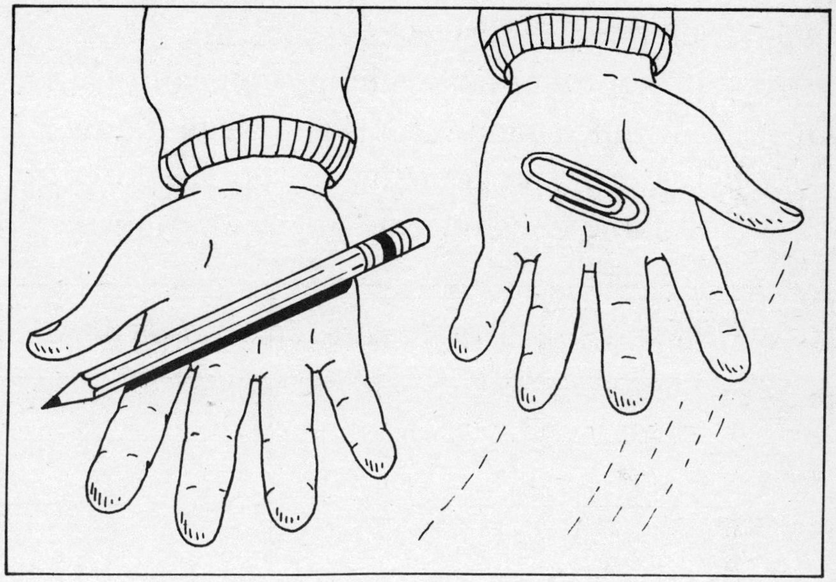

How much do you think a rubber band weighs?

Too Much
or Too Little?

You need: bathroom scales

Think about 5 objects in your house and how much each one weighs. (Don't pick the refrigerator or other heavy things you can't lift.) Guess how much each weighs and write it down. Put each of the objects on the scales. See how close you came to estimating the true weights.

Object	*Weight (est.)*	*Weight (lbs.)*

How close were your guesses? Try guessing your parents' and friends' weights and see how well you do.

Roll 'em

You need: a number cube, a pencil

- Sit down with your parent.
- Take turns rolling the number cube.
- Record the number that comes up each time you roll.
- Together, roll the number cube six times.
- On the seventh roll, try to predict which number will come up.
- Play the game four times.

1. ______
2. ______
3. ______
4. ______
5. ______
6. ______
7. PREDICTION ______

1. ______
2. ______
3. ______
4. ______
5. ______
6. ______
7. PREDICTION ______

1. ______
2. ______
3. ______
4. ______
5. ______
6. ______
7. PREDICTION ______

1. ______
2. ______
3. ______
4. ______
5. ______
6. ______
7. PREDICTION ______

Do you see a pattern developing?

Is there any one number that comes up most often?

Is there any way anybody could predict the next number and be right most of the time?

The Paper Clip Drop

Instructions:

- Color the boxes as directed in the illustration.
- Hold the paper clip or other small object about six inches above the center of the box.
- Drop the paper clip ten times.
- Record each color area that the paper clip hits on the tally sheet below.
- Predict which color it will hit on the eleventh hit. Explain why you made your prediction.

Ask your parent to do the same thing you did. Are the results the same? How are they different?

Tally Sheet	
Red	
Blue	
Yellow	

You need: blue, yellow, and red crayons, two paper clips or other small objects

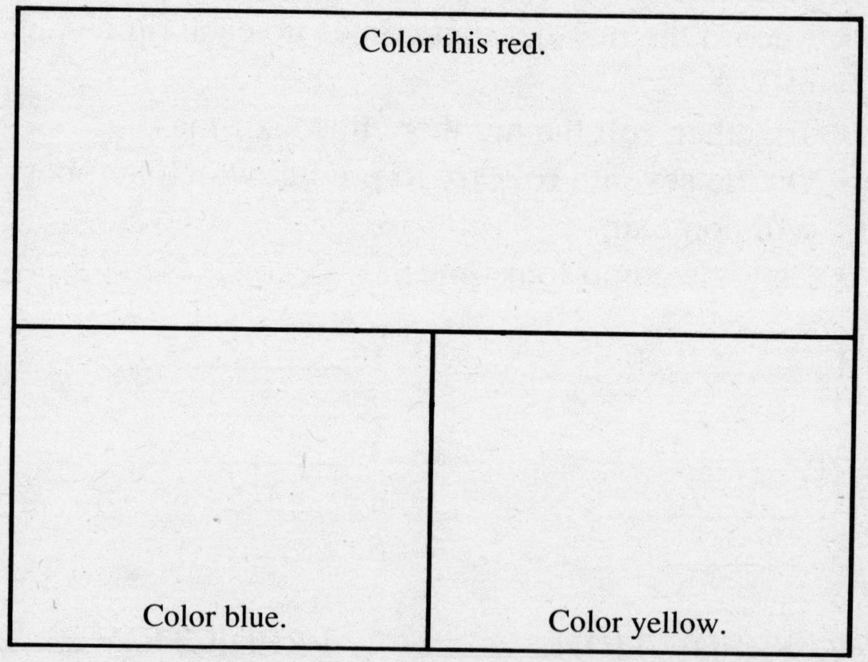

A Royal Family

You need: the jacks, queens, and kings from a deck of playing cards; paper and pencil

Lay out all the playing cards face up. Then answer the questions:

- How many different kings could the Queen of Hearts marry? ______
- If she could also marry the jacks (they're kind of like princes), how many different men could she marry? ______
- Could all the other queens also marry that many people, possibly? ______
- How many possible marriages, then, for all the queens put together? ______

Hint: Draw a picture or chart to help you.

CONCENTRATION CONCENTRATION CONCENTRATION

This will be a test of memory and skill at making predictions for you and your parent. Turn all of the cards face down on the flat surface. Shuffle the cards around and line them up on the surface. Then, to play:

- Take turns drawing the cards two at a time.
- The object is to find a matching pair. Once you have found a pair, keep those cards.
- Continue the game until all of the cards have been drawn.
- Whoever has the greater number of pairs is the winner.

Talk over these things:

- After you had drawn some of the cards, could you guess where the cards you needed were?
- Did you and your parent figure out where some of the cards were at the same time?
- Why do you think you were able to find the pairs more easily after some of the cards had been paired?

Play the game again, if your parent dares.

You need: all of the *red* numbered playing cards, 18 in all (no aces or face cards), a flat surface

The BUILDING Job

Extended Activity

Go for a walk with your parent outside to look at buildings. Take a notebook. Think about the plane and solid figures you see in the buildings. Do you find triangles in the rooflines? Are the buildings shaped like rectangular solids or cubes? What shapes are the windows and doors? Draw pictures of what you see. Think about how you could make little buildings out of boxes and paper. Draw the buildings you like. Tell why you like them.

YOUR OWN BUILDINGS (The Building Job, page 2)

Gather some small cardboard boxes to make a cluster of buildings. Make any group of buildings you choose, even fantasy buildings.

Choose a base for the buildings, such as a shoe box lid if your boxes are small, or the side of a cardboard box if your buildings will be larger. Ask your parent to cut windows and doors. Make roofs out of scrap cardboard. Glue the buildings together, if you like. Paint them if you can. Put curtains behind the windows.

Notice the solid and plane figures you are using to make your buildings. If you put in roads or paths, notice how you use points and line segments. What kinds of angles are in your buildings?

Name your buildings. Put up signs with the names on them.

A BUSY LITTLE CITY (The Building Job, page 3)

On this page, write a little story about the buildings you made. Pretend there are little people living in them. Tell what they are like, and what they do.

Create a Cone

Trace a circle like this around a saucer onto your paper, and cut it out. Find the approximate center and draw lines out to *A, B,* and *C,* as shown. Cut on the dark line from *A* to the center of the circle.

Put corner *A* over corner *B*. What happens to your cone if you move corner *A* over corner *C?* Staple or tape the corners together the way you like it.

Your cone could be many things. You could decorate it to be a tepee or a party hat. Think of what else your cone could be. Talk it over with your parent.

You need: paper, scissors, tape or staples, a saucer

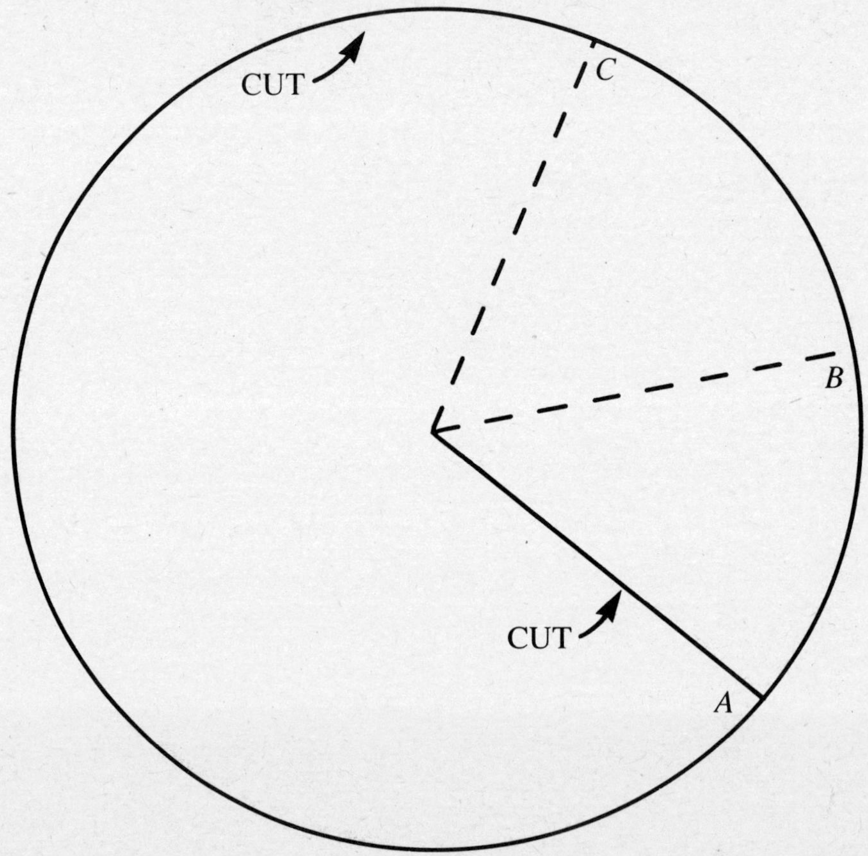

I Spy!

Play a game of I Spy! with your parent. You can be "It" and start the game. Look around for a solid figure such as those at the bottom of the page, but try not to let your parent know what you are looking at. Maybe you see a drink can. If you do, say "I spy a cylinder." Your parent must guess what cylinder you are thinking of. Next, your parent can spy something and you can guess what it is.

sphere cube cylinder cone rectangular solid pyramid

How many solid figures do you see out your back window?

Right-Angle Ruler

You need: a piece of cardboard, paper, scissors, glue or paste

Trace the right-angle ruler on this page. Cut out your tracing.

Paste or glue your tracing to a corner of the cardboard. Cut around the tracing.

Play a game with your parent. Take turns finding right angles in your house. Use your right-angle ruler to check.

How many right angles can you find in the room you are working in?

Plane-Figure People

You need: white paper, colored paper, a pencil, crayons, paste or glue

Trace the plane figures on this page.

Cut them out and trace around them to make more plane figures from your colored paper. You could even make some plane figures larger or smaller than your tracings.

Glue or paste your plane figures on the paper to make pictures of people and animals. You make some and let your parent make some.

Use line segments, points, and angles to complete your pictures. You might want to create some silly characters!

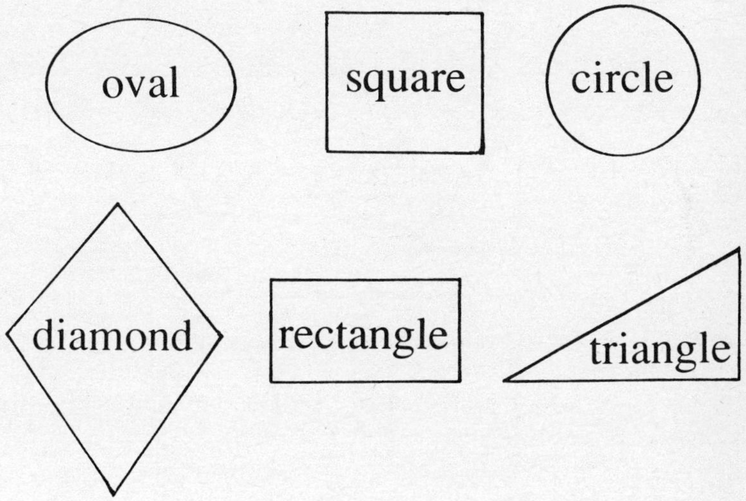

Congruent Memory

Trace and cut out two sets of each shape.

You need: paper, scissors, a pencil, glue or paste

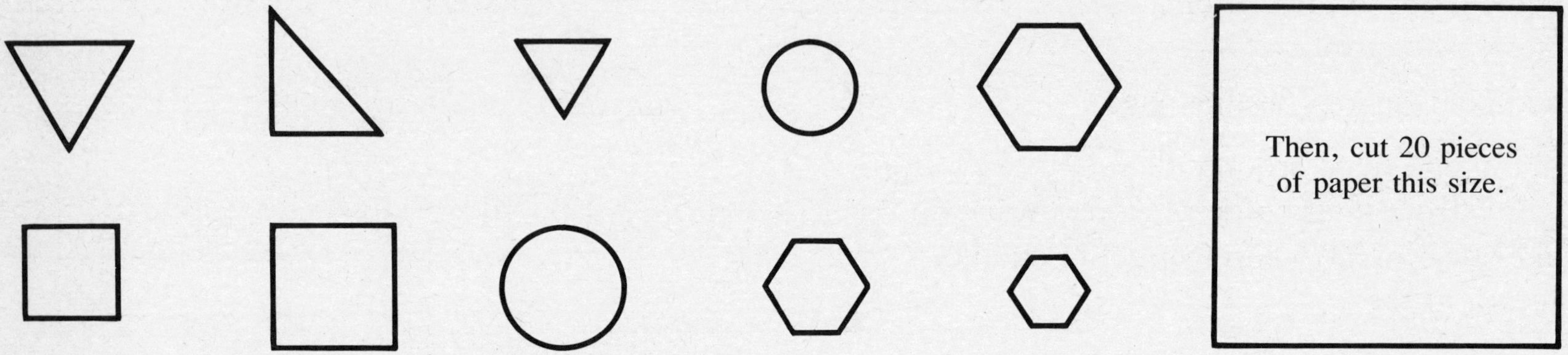

Paste one shape on each piece of paper to make cards. Play a game with your parent:

- Lay all the cards out face down. Take turns.
- Turn up two cards. If they show congruent shapes, keep them, and turn up two more cards.
- If they are not congruent, turn them upside down again, but try to remember where they were.

The game ends when all the pairs have been found. The one with more pairs wins.

Remember, congruent means having the same size and shape!

Decorate a Room

You need: scissors, colored paper, tape

It's easy for you and your parent to make paper decorations for your room or for a party. You can make them out of colored paper, like wrapping paper, the comic sections of old newspapers, or old magazines.

First, fold the paper in half. Start cutting from the fold and continue cutting until you cut across the fold again at a different place. Open the paper and you have made a design with symmetry. Put several colored designs together for decorations on a window or wall.

Try cutting these shapes.

A.

B.

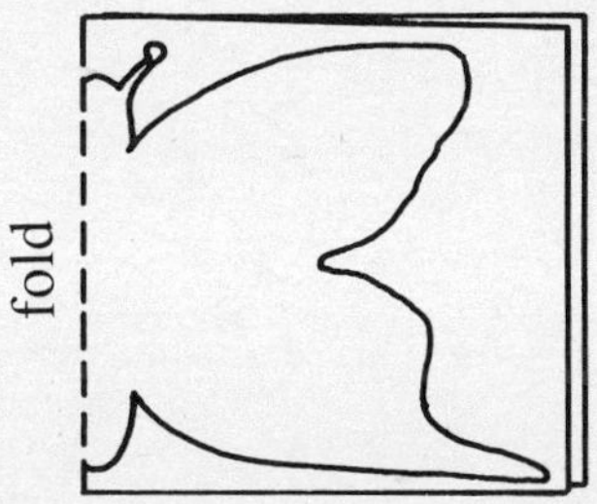

C.

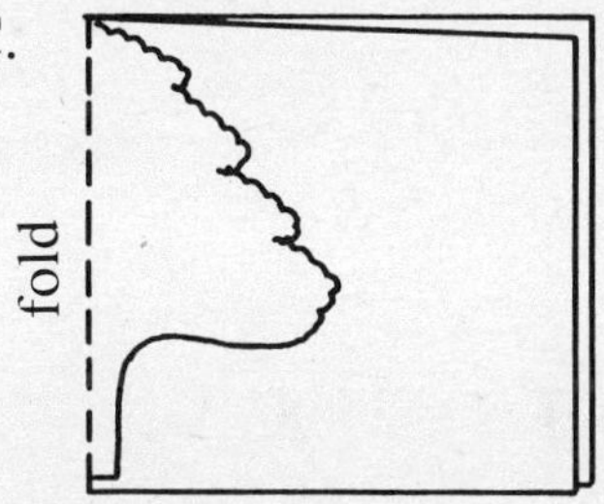

D.

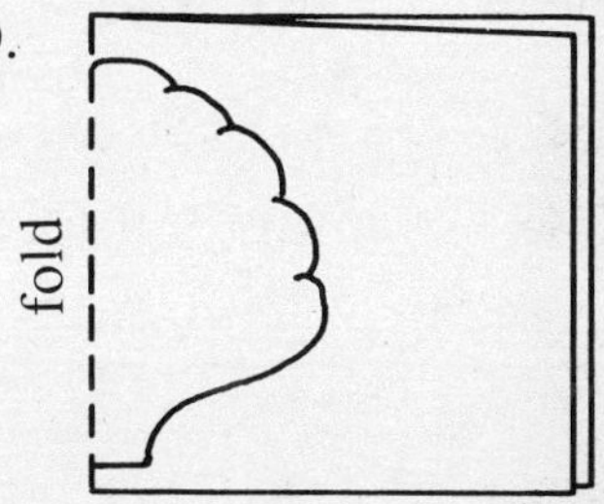

Make some symmetrical shapes of your own.

Peppy Placemats

Work with your parent to make placemats for your family. Each placemat needs to be about as big as two of these pages taped together, or a little larger.

Use only one shape on each placemat. Make a pattern with the shape. Try not to leave gaps in your pattern. Color your placemats.

You need: one big piece of paper for each placemat or two regular-size sheets, pencil, scissors, crayons, tape

Sample Placemat:

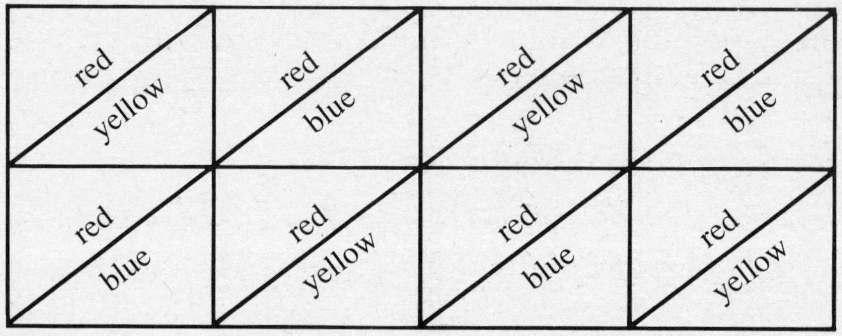

Trace these shapes and cut them out.

The SAME/DIFFERENT Game

Play a game of same and different with your parent. Use plane figures and solid figures from your home. Go get a bunch of things. Then, to play, take turns. On the first turn:

- Cover your eyes.
- Your parent groups 3 of the things together. Two of them must have something in common. (For example, it might be two boxes and a can of food.)
- You must tell which two are the same (the boxes, because they are rectangular solids—the can is different because it is a cylinder).
- Ask your parent to cover his or her eyes while you group the things together.

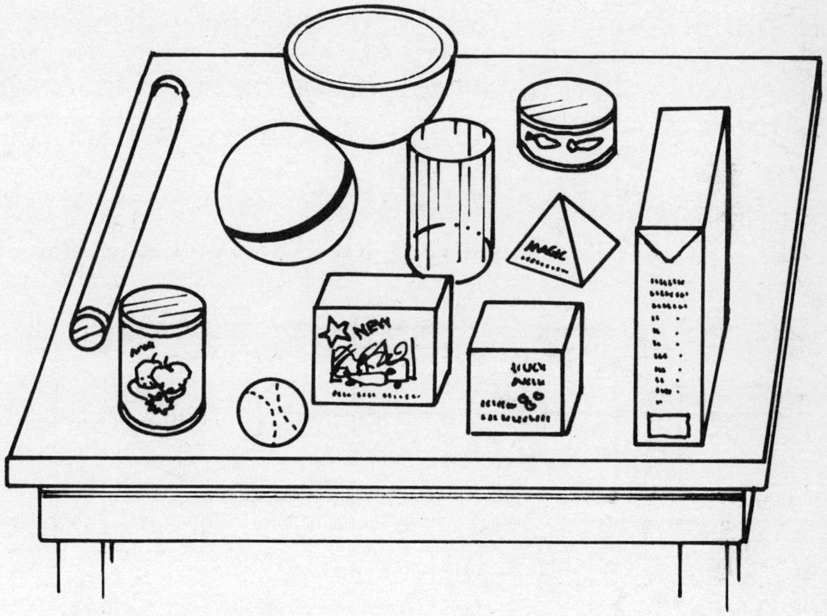

POINTS ON THE HOME GRID

Extended Activity

Think about a room in your house. Try to imagine making a map of it. Ask your parent to help you map this room, using squares, rectangles, ovals, and circles to represent the furniture. Draw in these shapes on the grid at the right.

Underneath the grid write out the name of the furniture and the grid points where it can be found. Using the points, have your parent call out the coordinates, and you try to find the piece of furniture being called out.

Below at left is a sample grid. Answer the questions about it before you begin your game.

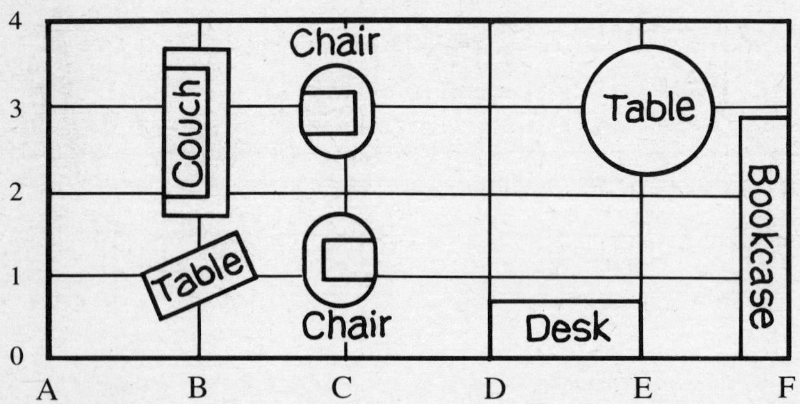

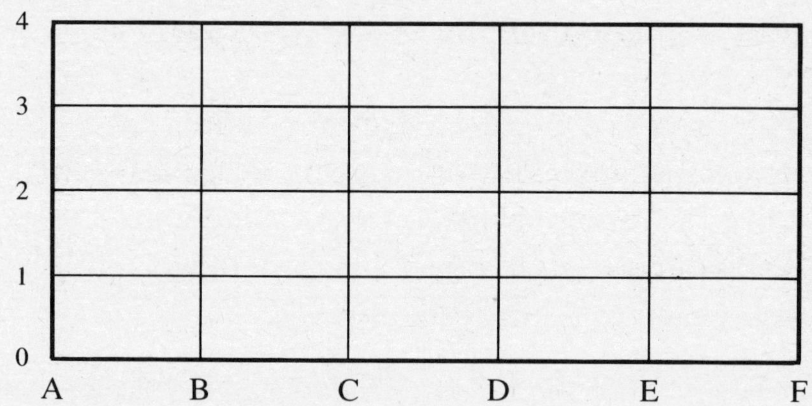

1. What is C3? ______________
2. What is E3? ______________
3. What is F1? ______________
4. What is B2? ______________

Furniture Name	*Coordinates*
______________	______________
______________	______________
______________	______________
______________	______________
______________	______________

FUN POINTS (Points on the Home Grid, page 2)

Do you want to play hide and seek? Your parent will try to find you using the map you will make on the grid.

- Map any other room on the grid. Place the large pieces of furniture on the grid, as well as any windows or doors.
- Decide where you might hide, and write down three possible places on a slip of paper, using the grid coordinates.
- Give the coordinates to your parent.
- Your parent will use the coordinates you marked on the grid to find you.
- Switch roles, and you look for your parent by grid coordinates after he or she hides.

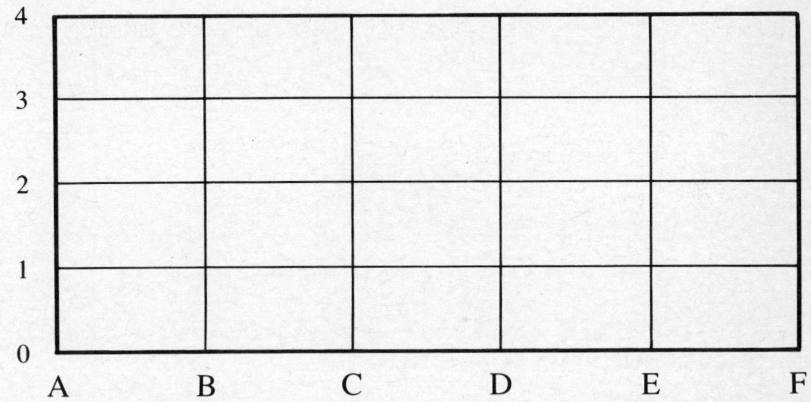

IMPORTANT POINTS (Points on the Home Grid, page 3)

Points on a grid can be used to map many things. These maps can be made of your whole house, to show all of the possible exits that could be used during a fire drill.

Use the grid at right to draw a map of your house. Use a red marker to show where all of the exits in the house are. Then using a blue marker, draw the quickest and shortest route from your room to the exits.

Mount your map on a piece of construction paper, and display it in your room. Now you and your parent can conduct a fire drill. Practice reading the coordinates for the exits.

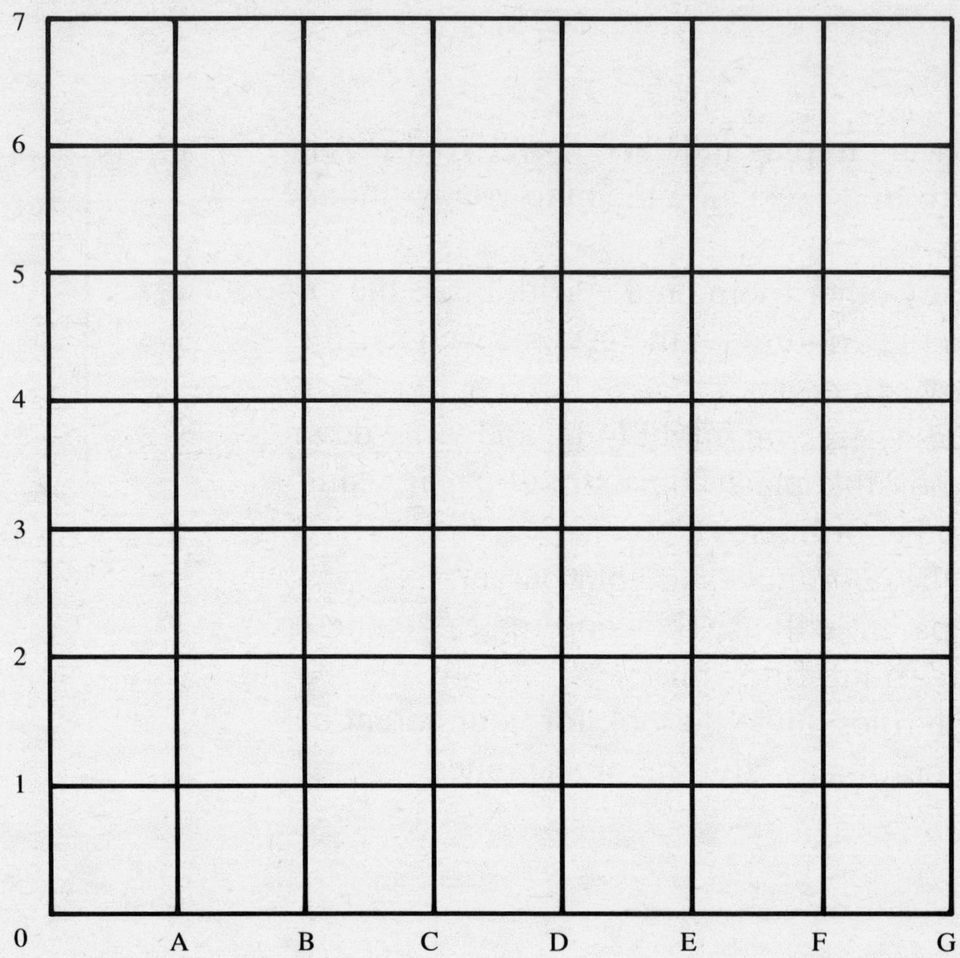

Part 6

More Multiplying and Dividing: you know how it's done

A note to parents

Mathematics is like exercise. There is no better way to learn what it's all about than by simply doing it. You can help your child get into "top mathematical shape" by being willing to do a few simple "exercises" with him or her each night.

Try the following suggestions. They will help you and your child warm up for the "main event" activities that come later.

- Let your child help you do some simple multiplication problems. For example, if each person in your family will eat 2 hot dogs at the picnic, how many hot dogs should you buy? (Multiply the number of family members by the number of hot dogs each will eat.)
- Show your child how to use division to determine a per-item cost. For example, if a bunch of bananas costs $1.10, how much does each banana cost? (Divide the total cost by the number of bananas.)

With a short "workout" each day, both you and your child will soon be in top mathematical shape!

Mental Math

Show your parent how you have learned to do mental math.

Use this problem:

> David is planting a garden. He wants to plant three rows of beans with 20 plants in each row. How many bean plants should he buy?

Show your parent how you think this problem through.

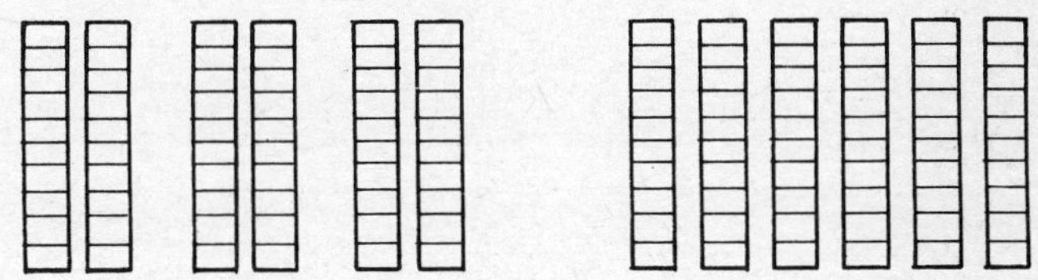

a. THINK: three rows of 20, so I multiply 3 times 20
$3 \times 20 = 3 \times 2$ tens
3×2 tens $= 6$ tens or 60

b. SAY: 3×20 is equal to 60. David needs 60 plants.

c. WRITE: $3 \times 20 = 60$.

You and your parent can make up story problems for many numbers. Take turns making up the story and finding the answer for these numbers, just by thinking.

$5 \times 40 =$	$7 \times 30 =$	$2 \times 90 =$
$3 \times 70 =$	$4 \times 50 =$	$5 \times 30 =$

Play this game in the car or when you and your parent are working together.

Fruit Shopping Spree

You need: paper, a pencil, scissors, crayons

Trace, color, and cut out each of the fruits on this page.

Put them on display in a pretend fruit stand.

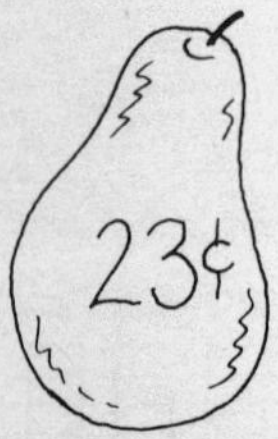

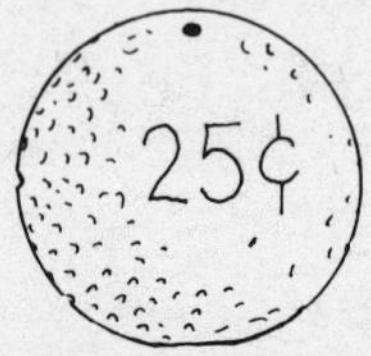

Play a game with your parent. Take turns shopping for fruit. Say, "I want to buy ________ (number) ________ (fruit). I estimate it will cost ____." On paper, show how you made your estimate. If your parent says you are close enough, take the fruit. If you are not close enough, in your parent's judgment, try again next turn.

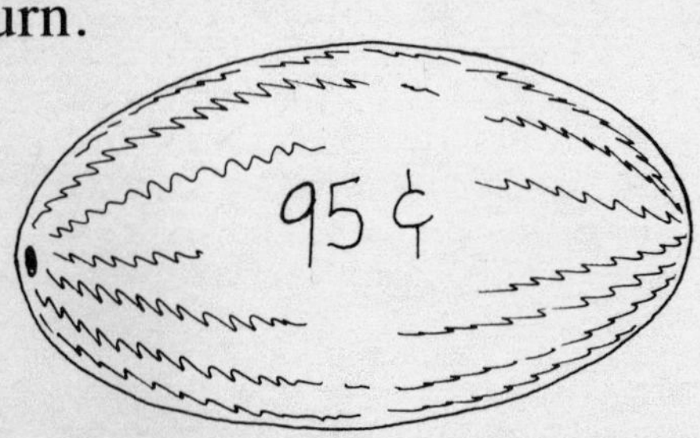

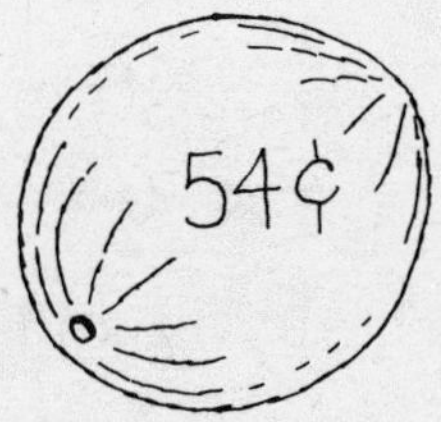

You could help your parent estimate the cost of fruits and vegetables when you shop for groceries.

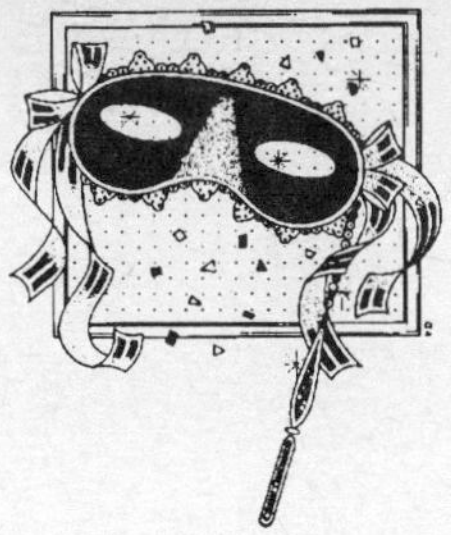

Party Plans

Ms. Sheridan and the parents are giving a class party, but they made some mistakes in shopping. Help them figure out what else they need to buy and what they should take back. There are 31 students in the class and 2 parents are helping. They have invited the principal, and they always take party food to the custodian. Ms. Sheridan thinks each person will drink one drink and eat 3 pieces of pizza, 1 apple, and 1 piece of cake. Here's what she bought:

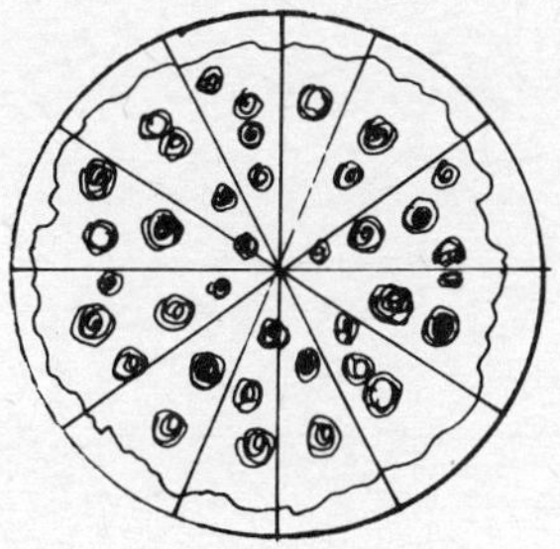

8 pizzas

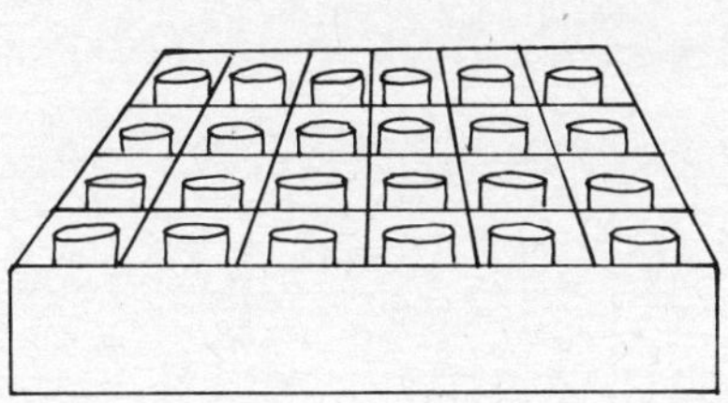

4 cases drinks

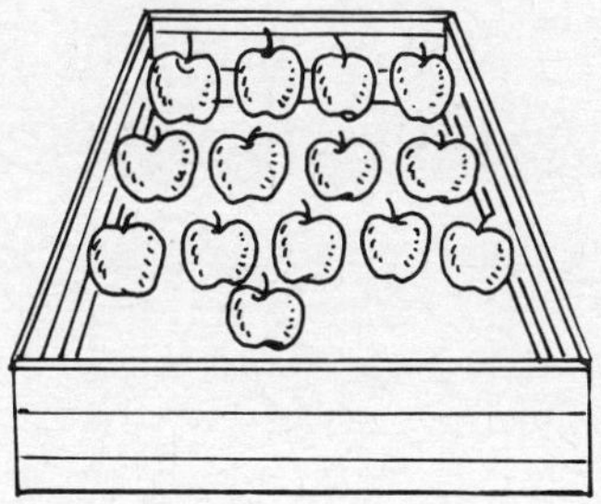

2 crates apples

3 cakes

What did Ms. Sheridan buy too much of? __________ How many does she really need? __________.

What will Ms. Sheridan need to buy more of? __________________________

How much more? __________

Remember: you cannot order only a part of some items.

SECRET Code

You need: paper and pencil

With your parent, write the alphabet in a column on a piece of paper. Then, starting with *A,* write numbers in numerical order beside each letter:

$A = 1$, $B = 2$, and so on.

Now work the problems below. Then look up the letter that matches each answer. Write the letter below the problem. When you're finished, you'll have a message!

$3 \times 6 =$ add 1	$2 \times 11 =$ add 1	$3 \times 3 =$	$2 \times 5 =$ add 3	$2 \times 6 =$ add 1	$2 \times 4 =$ add 1	$2 \times 7 =$	$2 \times 4 =$ subtract 1

$3 \times 4 =$ subtract 3	$2 \times 10 =$ subtract 1

$2 \times 12 =$ subtract 5	$2 \times 11 =$ subtract 1	$3 \times 4 =$ add 1	$2 \times 6 =$ add 1	$3 \times 5 =$ subtract 10	$4 \times 6 =$ subtract 6

$2 \times 3 =$	$3 \times 7 =$	$3 \times 7 =$ subtract 7

You could write secret messages to your friends this way!

Multiplier Maid

You need: scissors, pencil, light cardboard

Trace this shape and cut out 20 cards. (Or just make them about this size.) Use heavy paper or light cardboard. Then copy the problems on this page on the cards. Do not write the answers.

10	12	13	14	11
×2	×2	×2	×2	×3
10	11	10	11	10
×5	×5	×6	×7	×8
21	22	23	21	22
×3	×3	×3	×4	×4
30	31	32	41	43
×3	×3	×3	×2	×2

To play: Place the cards in a pile face down. Your parent picks up one card and shows it to you. If you answer the problem correctly, keep the card. See how quickly you can get all the cards.

Next time, try to do it even faster.

REGROUPING

You need: paper, pencil, and buttons, beans, or pebbles

Use a collection of small things—such as buttons, beans, or pebbles—to teach your parent how to regroup ones. Use the following problems:

15	13	15	17	16
×2	×5	×4	×3	×2

Now have a race with your parent to see who can do the following problems more quickly. Both of you should draw pictures of the ones to show how the ones are regrouped.

19	13	35	45	24	28
×2	×7	×2	×2	×3	×2

Step 1

Multiply the ones.

$$\begin{array}{r} \overset{1}{1}6 \\ \times\ \ 2 \\ \hline 2 \end{array}$$

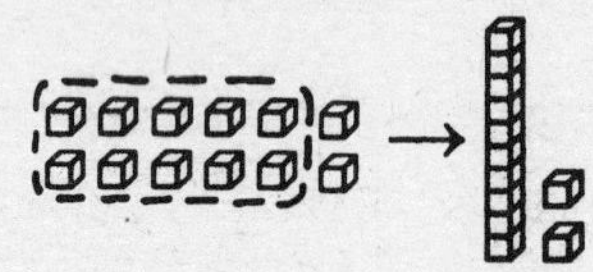

Rename 12 ones as 1 ten 2 ones.
Write 2 in the ones place.
Write 1 above the tens place.

Step 2

Multiply the tens.

$$\begin{array}{r} \overset{1}{1}6 \\ \times\ \ 2 \\ \hline 32 \end{array}$$

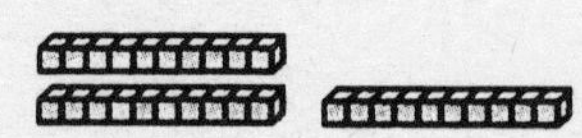

2 × 1 ten = 2 tens
Add the one ten.
Write 3 in the tens place.

You need: 2 crayons (different colors), a pencil, paper

The circus is in town. You and your parent want to buy balloons. To get a balloon, you must first work the problem on it. Take turns with your parent:

- Choose a balloon and work the problem.
- Your parent checks to see if the answer is right.
- If you're right, color the balloon in your color.
- Then your parent takes a turn.
- Check your parent's work.

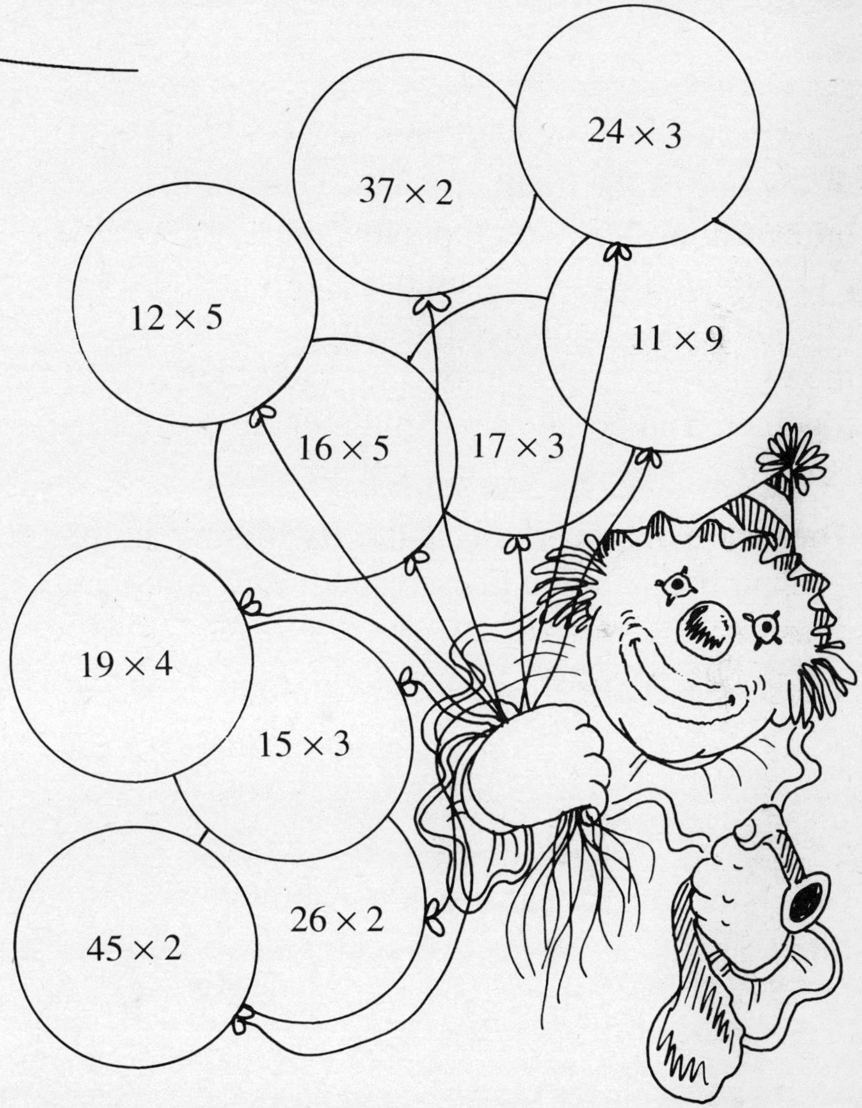

Secrets of a Number Cube

You need: 2 number cubes

Play this new game with your parent. Shake and throw one of the number cubes. Look at the number that shows on top. Now pick up the cube and look at the number on the bottom. Add those two numbers together. Throw it again and look at the top and bottom numbers and add them together. You get the same number, always! What is it?

Throw the number cube again. By looking at the top number, can you guess what the bottom number is?

Now, throw both cubes. Figure out what the bottom number is for each cube and multiply the bottom numbers together. (Do it in your head or use a piece of paper.)

Check the cubes, and if you get it right, you get to throw again. If you get it wrong, your parent throws.

Did your parent know the secret of the number cube?

HIGH LOW

With your parent's help, find your number cards. Put all the odd numbers in one cup and all the even numbers in the other. Then, to play:

- Take one card from each cup. Multiply the two numbers.
- Next, your parent takes a card from each cup and holds them up for you to see. Multiply those numbers together.
- Are you holding the higher answer or the lower answer? Say "High!" or "Low!"
- Now that you know the rules, try to quickly *estimate* if you have the high or the low number, and say it. You don't need to work out the exact answer.
- If you are right, you get to keep all the cards. If you are wrong, you have to put them back in the cups and mix them up.

Were you able to get all the cards?

You need: number cards 1-10 from the back of the book, 2 cups

Grocery Shopping

Make a trip to the grocery store with your parent. Notice how many times you need to multiply.

First, check off any items you needed, on the chart on the following page. The chart has some blank spaces for you to write in some other food you need to buy.

When you've finished, take the chart home. You and your parent talk about the following questions:

1. How much did you spend on your food?
2. Were you surprised at the cost?
3. Did you waste any money, in your parent's opinion?

MAKE YOUR LIST (Grocery Shopping, page 2)

You and your parent take this chart to the grocery store. You fill in the blanks you can.

ITEM	*NUMBER OF POUNDS OR NUMBER OF ITEMS*	*COST PER POUND OR COST PER ITEM*	*TOTAL COST*
apples			
oranges			
bananas			
lettuce			
potatoes			
carrots			
onions			
chicken			
fish			
dry cereal			
milk			
cottage cheese			
yogurt			
beans			
rice			
bread			

WHY MULTIPLY? (Grocery Shopping, page 3)

In this space, explain how multiplying is important in shopping.

Eating In Eating Out

Get the receipt from your family's last shopping trip to the grocery store.

With your parent, figure out how much each item of tonight's dinner cost.

Add to find the total cost of tonight's dinner.

Then divide the cost by the number of people at dinner.

How much did each person's meal cost?

Ask your parent how much dinner for one person costs at a typical restaurant.

Do you think it's a better idea to eat at home or eat out often? Why?

You need: a receipt from the grocery store

Oh Those Odometers!

With your parent, read the mileage on the odometer of your car or someone's car. Read only the whole miles.

The odometer tells you how many miles the car has been driven since it was new.

Now, find out what year the car was sold new. Figure out how many years old it is.

Divide age into miles. This will tell you about how many miles the car was driven—on the average—every year.

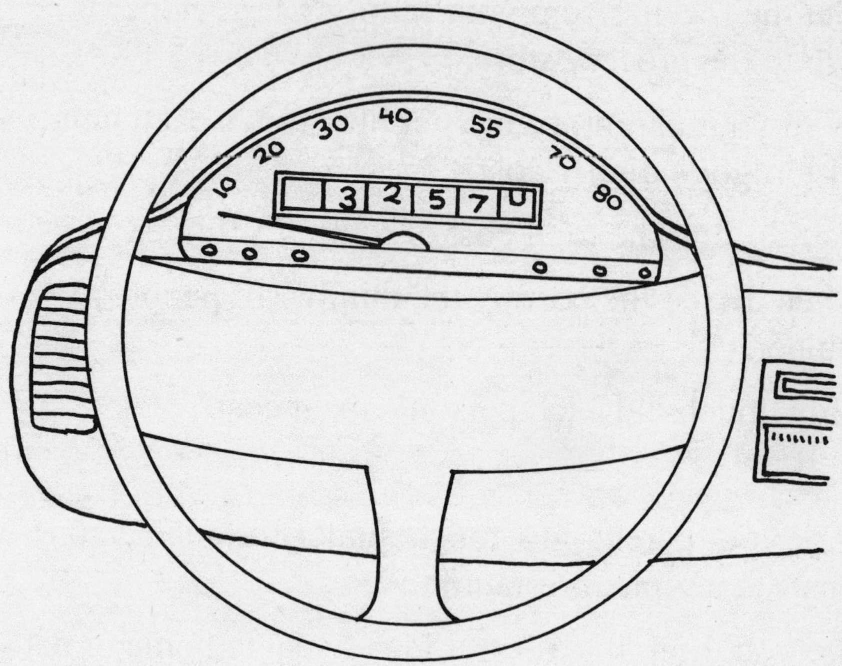

Do the same for another car. How do the numbers compare? Why might one car be driven more in a year than another?

CUTOUT PAGE (Number Squares, Centimeter Ruler)

Be sure to keep your number squares and your ruler in a safe place, to use next time.

1	2	3	4	5	6	7	8	9	10
11	12	13	14	15	16	17	18	19	20
1	2	3	4	5	6	7	8	9	10
11	12	13	14	15	16	17	18	19	20
1	2	3	4	5	6	7	8	9	10

Centimeter Ruler

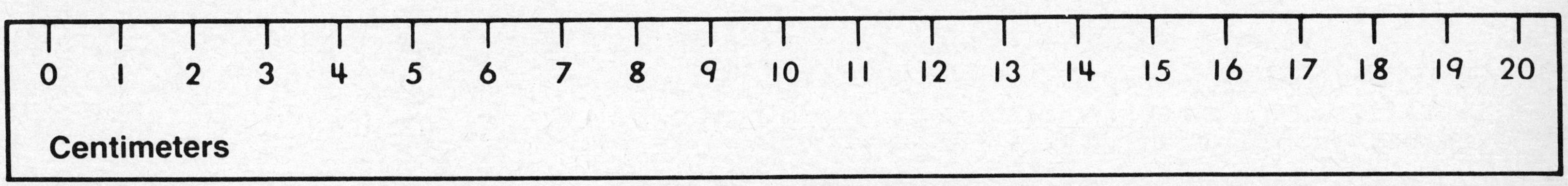

CUTOUT PAGE (Money)

Reminder: Be sure to keep your cutout money someplace safe, like in a little margarine tub, for next time.

CUTOUT PAGE (Money, page 2)

Reminder: Be sure to keep your cutout money someplace safe, like in a little margarine tub, for next time.

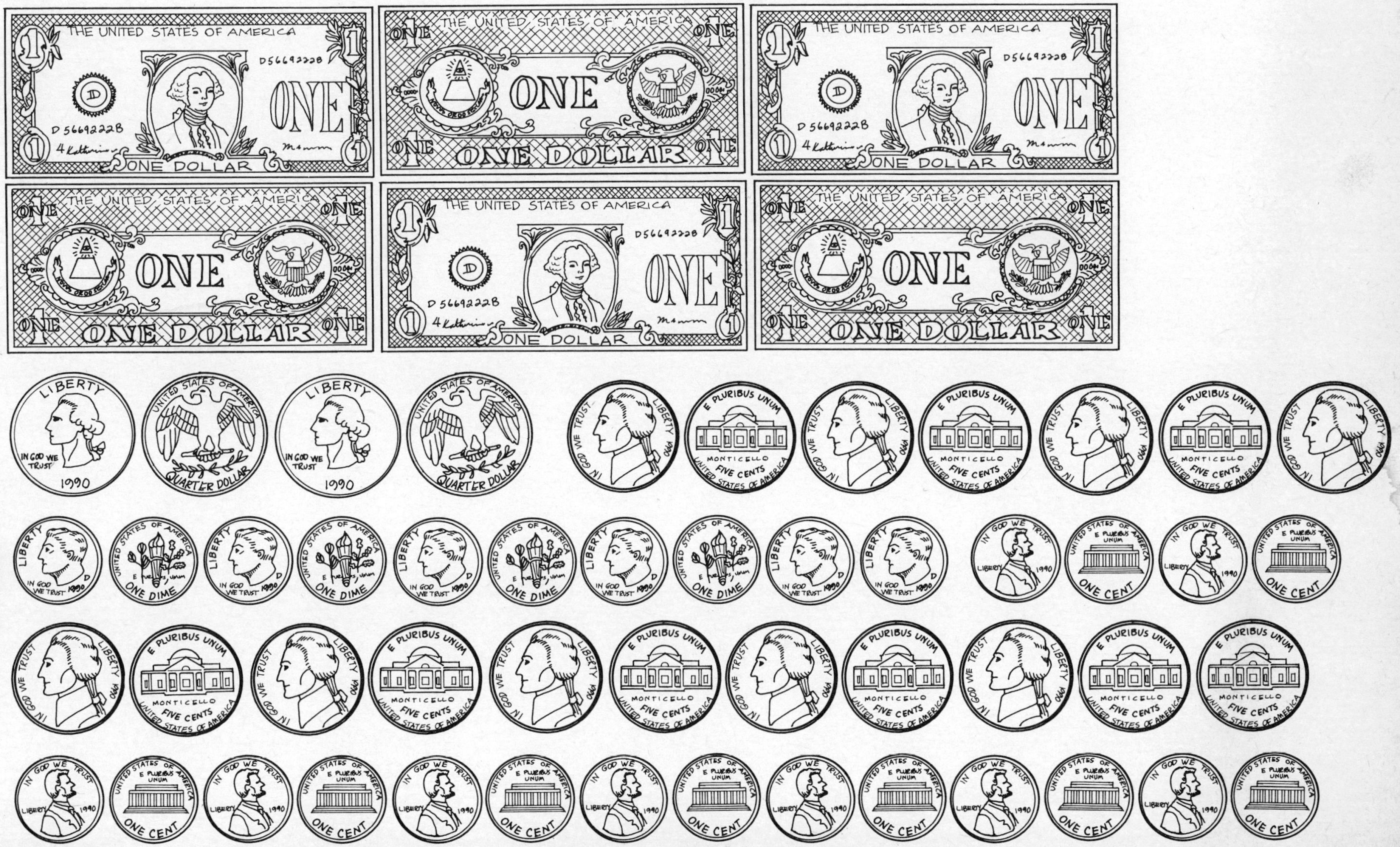